THE HISTORY OF
Topographical Maps

P. D. A. Harvey

THE HISTORY OF
Topographical Maps

SYMBOLS, PICTURES
AND SURVEYS

with 116 illustrations, 10 in color

Thames and Hudson

© 1980 Thames and Hudson Ltd, London

First published in the USA in 1980 by
Thames and Hudson Inc., 500 Fifth
Avenue, New York, New York 10036.

Any copy of this book issued by the
publisher as a paperback is sold
subject to the condition that it shall
not, by way of trade or otherwise, be
lent, resold, hired out or otherwise
circulated, without the publisher's
prior consent, in any form of binding
or cover other than that in which it is
published, and without a similar
condition including this condition
being imposed on a subsequent
purchaser.

All rights reserved. No part of this
publication may be reproduced or
transmitted in any form or by any
means, electronic or mechanical,
including photocopy, recording or any
information storage and retrieval
system without permission in writing
from the publisher.

Library of Congress Catalog Card
Number: 80-80086

Color illustrations originated in
Switzerland by Cliché Lux and
printed in Great Britain by George
Over Limited, Rugby, Warwickshire.
Text and monochrome illustrations
printed in Great Britain by BAS
Printers Limited, Over Wallop, Hampshire.
Bound in Great Britain by
Webb, Son & Co. Ltd, Glamorgan.

Contents

Preface

THIS BOOK is an answer to the question that occurred to me twenty-two years ago when I first saw the superb sixteenth-century maps collected by Sir Robert Cotton and now in the British Library.

'They aren't maps at all,' I wanted to say; 'they're pictures, bird's-eye views. Why are they called maps?'

And despite continuing work on the history of mapping, despite increasing knowledge of the topographical maps of the sixteenth and seventeenth centuries, I never found a book that told me why.

So I set about writing it myself and the present book is the result. It is addressed to two audiences. To the non-specialist general reader it introduces a range of early maps that are scantily treated in most books on cartographical history. A few of these maps are beautiful, many are picturesque and all are interesting, not least because they show so clearly how, in many different societies and many different ages, men have found their way, slowly and hesitantly, towards that particular method of representing landscape that we call the map. But I hope it may have something to say too to the specialist historian of cartography: it is a first attempt to interpret and analyse a type of cartographic material that is often regarded – wrongly, I believe – as having little to do with the development of the maps we use today.

Because the book is essentially an essay in interpretation, not a work of primary research, it seems inappropriate to fortify it with elaborate references to justify every statement it contains; but because it draws on many sources of information, some far from obvious, a note on authorities is appended to each chapter. I hope these notes will meet the needs of specialist and non-specialist reader alike; they are meant to take the place of footnote references, not of a full or select bibliography. A glance at any of these notes will show that many of the maps discussed here have hitherto been considered only in a local or, at best, national context, not as part of the general history of cartography. One result is that language has posed a real problem. I have been notably hampered by my ignorance of, in particular, Arabic, Chinese, Japanese and Russian, and I thus may well have failed to do justice to early achievements in local mapping in the relevant cultural areas.

A history of cartography can all too easily become little more than a catalogue, a set of descriptions of one map after another. I have tried to avoid this by concentrating on general problems rather than on individual maps. There is no lack of these problems and on nearly every page unanswered questions arise – unanswered but not unanswerable: some might be answered by a week or so's work in a major library, others demand the time and scope of a doctoral thesis for their solution. I hope that by drawing attention to the problems I have met I may arouse other people's interest or curiosity to investigate them further.

Work for the book has placed me in many people's debt. I particularly owe thanks to many historians of cartography who in conversations, in lectures or in their writings have given me information or ideas that would otherwise have eluded me. I am especially grateful to Dr J. B. Harley, whose advice and encouragement were of particular help in the earliest stages of planning the book, and to Dr Juergen Schulz, to whose knowledge and kindness I owe far more than might appear even from the extensive reference I have made to his important work on De' Barbari's view of Venice. I am grateful to Professor J. S. Bromley, Mr A. C. Duke, Professor C. Morris and Mr T. B. Pugh, my former colleagues at the University of Southampton, for invaluable references to specialist writings. And I am grateful too to the Delegates of the Oxford University Press for permission to use material gathered for the forthcoming *Local Maps and Plans from Medieval England*, edited by the late Dr R. A. Skelton and myself; and to Dr Joseph Needham and the Cambridge University Press for permission to quote his translations from Chinese authors in *Science and Civilisation in China*, vol. iii. The book could never have been written without the courteous help of the staffs of the libraries where most of the reading for it was done: Universitetsbiblioteket at Uppsala, Kungliga Biblioteket at Stockholm, the British Library Reference Division (both the Reading Room and the Map Library) and the University Libraries of Southampton and Durham (particularly their Inter-Library Loan sections). My wife has not only patiently put up with work that has dominated my thought and conversation for the past year, but has also, by reading the whole book in manuscript, eliminated much clumsiness of expression.

University of Durham,   *P. D. A. Harvey*
12 July 1979

Introduction

THIS BOOK is about the early development of topographical mapping. By a topographical map we mean a large-scale map, one that sets out to convey the shape and pattern of landscape, showing a tiny portion of the earth's surface as it lies within one's own direct experience, and quite distinct from the small-scale maps that show us the features of whole provinces, nations and continents. At first sight this distinction may seem arbitrary and meaningless. We are used to being able to consult maps on a great variety of scales, covering larger or smaller areas in many different degrees of detail – and this has been the case in Europe for the last four hundred years. Where and why should we draw a line to separate these so-called topographical maps from the rest? But if we move to earlier periods or other societies we find the distinction far clearer. The second-century geographer Ptolemy (Claudius Ptolemaeus) distinguishes between 'geographic maps' of the whole world, which show features by lines and dots, and 'chorographic maps' of smaller areas, which make use of some pictorial elements. His chorographic maps are simply topographical maps – indeed the two terms are still used synonymously along with a third, 'cadastral maps' (which is perhaps more usefully restricted to its original meaning of an official map drawn to serve as a basis for taxation). To Ptolemy the difference between these and the geographic maps was fundamental, and it in fact reflects their completely different origins and early history. Occasionally we shall find the distinction becoming a little blurred, as in some medieval maps of Palestine or parts of Italy, and at one point in our story (chapter 8) we shall ignore it altogether, for the history of the itinerary map involves both sorts of mapping. But for the most part the distinction

is clear; our theme is neither an arbitrary nor an artificial one.

Let us start by looking at a modern topographical map, the official Ordnance Survey map of Great Britain on the scale of one inch to a mile (1:63360). The small portion reproduced here shows the city of Canterbury and an area to the north and east; the map was published in 1959. In ills 2 and 3 we see air photographs of the same area. It is obvious at once that each gives some information that the other does not. The map shows us some things that are not actually visible in the landscape at all: place-names, contour lines, administrative boundaries, the kilometre squares of the reference grid. Equally the photograph shows us some things that do not appear on the map, which is selective in the detail it includes and leaves out, for instance, field boundaries and individual houses in built-up areas. But apart from these differences in the information they provide there are two fundamental differences between the photograph and the map. One is that even the so-called vertical air photograph shows the landscape from directly above only at a single point immediately below the camera; elsewhere its view is more or less oblique with corresponding differences (however minute) in shape and scale. The map, on the other hand, achieves the visually impossible feat of vertical representation at every point, to a uniform scale throughout. The second difference is that the map, instead of reproducing the actual shape, colour and texture of the features it shows, replaces them with a set of more or less elaborate signs and symbols by which they are displayed far more clearly and far more uniformly than they appear on the photograph. When an eighteenth-century household inventory referred to 'One Mapp & Eight ordinary Picktures' it was not far wide of the mark. A topographical map is a picture, but it is a picture of a very peculiar sort: it represents the features of the landscape by conventional signs instead of their real appearance, and it sets them in a framework constructed by scrupulously careful surveying to attain an exactly uniform scale over every inch of the map's surface.

Looking at it this way we see that our topographical map, which at first sight seems so simple and natural a representation of landscape, is in fact a highly artificial construction. We are so accustomed to maps of this sort, to using them every day of our lives, that it can be quite hard for us to realize that the concepts they embody are very sophisticated and that it is only to be expected that the topographical map developed at quite a late stage in man's cultural history. There is no word for map in any ancient European language. In some modern European languages, such as Spanish, English and Polish, the word used derives from late Latin *mappa*, a cloth, by way of *mappa mundi*, a cloth painted with a representation of the world. In most others it derives from late Latin *carta*, which meant any sort of formal document, resulting in ambiguities that persist to this day; thus French *carte*, Italian *carta*, Russian *kárta*, can all mean other things besides a map. We find a similar pattern in non-European languages. In most Indian languages the word for map derives from the Arabic *naqshah*, but other meanings attaching to it include picture, general description, or even official report. In Chinese, *thu* is no less ambiguous: besides map it can also mean a drawing or diagram of any kind. All this reflects the fact that the map is a relative latecomer to the cultural scene. The name applied to it was that of a closely related object, and this name may or may not have acquired a more specialized, more specific, meaning as the map became fully established as an object in its own right. We have very few topographical maps from medieval England, and fewer still on which their authors refer to their own products; when they do so they are clearly at a loss to know what to call them, but opt for some Latin word meaning simply illustration – *pictura*, *figura* or *effigies*.

That this difficulty should have arisen at all, and that these were the words chosen to solve it, are matters of some significance in the history of the topographical map. But they concern a part

of its history that has been little written about. If we look at a modern history of cartography – and there are some very good ones – or at articles and monographs on particular aspects of the history of mapping, we have no trouble in tracing the development of our topographical map back to the early sixteenth century. It has come a long way since then, and we will find the principal landmarks on its journey fully investigated and charted: the advances in cartography and surveying, such as the invention of contour lines and the use of aerial photogrammetry, the great achievements, such as the survey of Bavaria by Philipp Apianus (1554–61), the mapping of the Scottish Highlands under David Watson and William Roy (1747–54), or the Cassini family's map of France (completed in 1793), all these belong to a history that is well known and is clear in its outlines. Just how far the topographical map has come since the sixteenth century we can see by comparing our modern topographical map with the map of the Black Forest by Johann Georg Tibianus, first published in 1578 and reissued in

1

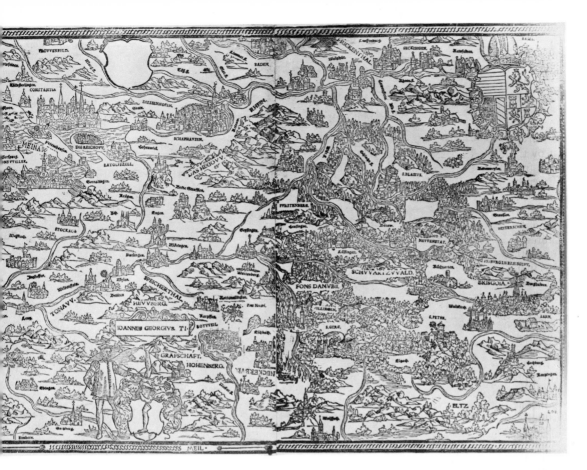

1603. The differences are plain to the eye, but in a sense the developments that have occurred are little more than refinements. In its essentials the topographical map was already complete by the early sixteenth century. The concept – the highly sophisticated concept – of a landscape portrayed by selected features shown in a standardized form and set in the framework of a uniform scale underlies the map of Tibianus just as it underlies our modern map of the area around Canterbury.

It is when we try to probe deeper into the genesis of our topographical map that our guides fail us. Certainly they will tell us all that is known – and it is a great deal – about the earlier history of smaller-scale maps of countries and continents, what Ptolemy called geographic maps, and of the growth of knowledge of the earth's form and surface that lay behind them. The ancient Greeks had proceeded a long way towards a true theory of geography and this, allied to the knowledge of distant lands acquired by military expansion and commerce in the Roman Empire, culminated in the work of Ptolemy himself: a treatise on geography accom-

1 Map of the Black Forest, facing south, by Johann Georg Tibianus, published 1578 and reissued 1603. Pictures take the place of conventional signs except for the abbot's crozier marking monasteries.

panied by a set of maps of the world as then known in Europe, roughly speaking the northern hemisphere from Japan to the British Isles. Ptolemy's maps are drawn with a grid of latitude and longitude (the former fairly correct, the latter less so), and their general accuracy of outline is startling, especially when we remember that they have come down to us in copies made a thousand years or more after Ptolemy's own time and probably at many removes from his original maps. But Ptolemy's work survived only in Greek and Arab traditions, and the world view of western Christendom in the middle ages produced maps of a very different sort which may ultimately have derived from Babylonian sources. Circular in shape, with Jerusalem at the centre, the simplest of them show no more than the division into the three continents by means of a T representing the Mediterranean and the Red Sea, while the more elaborate show countries and oceans in a jumbled profusion that owes little to their real shapes and relationships – the late-thirteenth-century map by Richard of Haldingham in Hereford Cathedral is an outstanding example. But by the later middle ages there were other, more fruitful, developments. From the late thirteenth century, navigators in the Mediterranean were making use of maps, the so-called portolan charts, with reasonably accurate outlines of the coasts and naming ports and other features; and the increasing knowledge that came with longer voyages was reflected on the maps in improved and more extended outlines, first along the English Channel and North Sea and then, in the fifteenth century, along the north-west African coast. Meanwhile Ptolemy's work had been discovered by the west: a Latin translation, completed by 1406, achieved rapid popularity, so that it survives today in nearly fifty manuscripts of its original version as well as in copies of various improved versions from the 1460s on and in a series of printed editions, the first as early as 1475. And a new interest in theoretical geography, particularly in central Europe, led to the compilation of position tables giving the latitude and longitude of lists of towns, at least by the 1440s and perhaps some twenty years earlier, and led too to the construction of some maps on this basis. This European tradition of geographical theory and of mapping large areas was not the only one in the contemporary world; the Arabs and the Chinese had traditions of their own. But it was the European tradition which lay behind the geographical discoveries and the small-scale maps of the sixteenth century, and thus came to form the basis of modern geography.

But all this has little to do with topographical mapping. All that can be said is that working out geographical coordinates of places enabled the small areas shown on large-scale maps to be correctly placed in their wider context. And when we try to discover what was happening in topographical mapping before the sixteenth century, what line of development lay behind the maps like Tibianus's, we are offered only isolated examples of earlier maps of this sort, or even mere guesses and assumptions that at best fail to add up to a coherent pattern. 'We

know that in Eratosthenes' time [i.e. the third century BC] vast cadastral surveys of the Nile lands had been made. It is likely that there were some fairly good maps of that region.' 'The Greeks were the first to lay out complete cities according to a well thought out plan. Their great townplanners (such as Hippodamos of Milet in the fifth century B.C.) must have drawn this plan in advance.' 'One might suppose and suggest that William, the Norman, had special maps of his Conquest. . . . His work Domesday Book gave a detailed description of each district, whether cultivated, settled or uninhabited. These details put him in a position to make topographical maps.' In compiling (in the seventeenth century) a map showing North Friesland as it had been in 1240, 'Even for this period older plans were available'. In fifteenth-century England 'small local maps were commonplace, made by surveyors using "new" continental methods to map towns or estates involved in property disputes'. All these statements about early topographical maps are taken from the writings of cartographic historians of repute and distinction; not one can be substantiated and some are demonstrably untrue. Perhaps the historians of maps have become victims of their own expertise. To compile a map of the world, or even of a single kingdom, with any accuracy calls for such skill and knowledge of geographical principles that it can only be seen as the result of a long process of thought and discovery; on the other hand the topographical map, in the form we find in the sixteenth century, seems so natural and straightforward a production that it calls for no particular explanation, no need to look for its origins. While the maps of certain primitive peoples, such as the Eskimos or the Marshall Islanders, have been the subject of interest and comment, it has often been taken for granted that more advanced societies will have been as familiar with topographical maps as we are ourselves. Indeed some historians of cartography draw particular attention to maps made in very distant ages or in very primitive societies with the clear implication that drawing maps comes so naturally to people of all ages and cultures that it is practically instinctive. And given the ambiguity of the word used for map in many languages it becomes very easy to find maps where none ever were; a case in point is the map of the land of Canaan that has been postulated on the basis of the eighteenth chapter of the Book of Joshua.

But there can, after all, be few people who are quite as familiar with the idea of drawing maps as historians of cartography; and, as we have suggested, the concepts behind the natural, straightforward topographical map are very far from simple ones. Their growth is as much a part of the history of cartography as the theories and discoveries that led to man's ability to map large areas. The aim of this book is to fill this gap in the history of map-making and to show just what did lie behind the topographical maps like that of Tibianus that were produced in the sixteenth century. At best it can only be a first essay. Because early topographical maps have never been systematically studied as a whole there are many places where the story is incomplete, where we have much more to learn. But we can at least see how far our present knowledge falls into a pattern, a pattern which future research and discoveries may elaborate or correct. In fact in broadest outline the early development of topographical mapping seems both very clear and very simple, and it is summed up in our book's sub-title: it is a development from symbols to pictures, and from pictures to surveys.

Let us turn back to our map of the area around Canterbury. As we have seen, one way it differs from an air photograph is in representing features of landscape by conventional signs, not by their actual appearance. Some of these signs have no pictorial significance at all, like the red triangle that marks a youth hostel two miles north of Canterbury Cathedral or the black lines and red circles that mark the railway and stations. But one reason why it is not difficult to see the map as a special kind of landscape picture is that so many of the conventional signs are pictorial ones. They are not actual pictures of the individual features they represent, but they are stylized pictures of typical examples of those features. Often the pictorial element is very obvious. Woodlands appear as patches of green dotted about with trees whose shape shows whether they are mostly deciduous, coniferous or mixed; orchards are marked by smaller, neater trees set out in regular rows; the rivers, streams and ponds are coloured blue; by Hoplands Farm (near Westbere) black dots and

tiny circles represent stones in the open pits of a mine. Sometimes the pictorial origin of a sign could be overlooked: the blue crosses that mark glasshouses on the north-west side of Canterbury, the sets of horizontal dots showing that an area north-east of the town is rough pasture, the fact that roads are shown either in brown or in the closely related red or yellow. Whether it at once leaps to the eye or not, a great many of the conventional signs used on this particular map are based on the actual appearance of the feature, viewed from above (like the glasshouses) or from the side (like the trees in the woods and the orchards).

Now let us look at another modern map, a portion of one sheet, published in 1944, of an official map of Switzerland. It shows a landscape $\quad$ II of a very different sort, the area between Zermatt and the Matterhorn, high up in the Alps on the Swiss-Italian border. It is on nearly the same scale as the British map (1 : 50000 instead of 1 : 63360), and the portion shown thus covers about the same area of ground, though the terrain is so different that it takes a little effort to appreciate this. Both maps are the products of long traditions of topographical mapping in their own countries. Both are extraordinarily beautiful maps: they are aesthetically pleasing to layman and professional cartographer alike. At first sight the Swiss map seems far more pictorial than the British one. In point of fact it gives much less information about the human landscape, and for the features it shows – houses, railways, cultivated areas – it uses signs that are not pictorial. But the map is entirely dominated, as the landscape is entirely dominated, by the mountains and glaciers, and these are shown not by contour lines alone (as the relief is shown on the British map) but by an elaborate and skilful hachuring and other markings which bring the whole area to life before us, the high peaks, the rock faces, the mountain shoulders and ridges, the cliffs of ice, all standing up from the map as though in three-dimensional form. At the same time it is not a straightforward picture of the high mountains; cartographic convention comes between their actual appearance and the map, so that, for instance, the hachuring is drawn to give the appearance of light coming from the top left corner, that is from the north-west, a direction from which the sun never shines in the northern

13

hemisphere. As one geographer put it, referring to this style of hachuring, 'The basis for the representation of relief is not the laws of physical optics, but rather ... the laws of cartographic optics'.

But even in the late nineteenth century some cartographers held that the convention of north-west illumination was unacceptable because it could not occur in reality. In fact one of the most significant changes to occur in topographical maps since the sixteenth century is the steady replacement of simple pictures of landscape by signs and conventions that may or may not have a pictorial origin. The pictorial elements in our modern British and Swiss maps are surviving vestiges of a time when maps were wholly pictorial. Let us look again at the map of the Black Forest by Tibianus. It marks abbeys by the head of a crozier drawn above the buildings, but otherwise, without being in any sense a simple view of the landscape, it consists entirely of landscape pictures: no key is needed to tell us what is a mountain, what is a wood or what is a town, and the towns themselves are shown not by a single uniform sign but by tiny thumb-nail sketches that are at least related to the actual appearance of the particular places. In this the map is like many topographical maps of its time. On the other hand we can also see that it has taken the first steps towards standardization: the pictures of towns, for instance, while differing in their features are fairly uniform in size apart from the large cities, Constance, Schaffhausen and one or two others.

But in another way too we can see that the map of Tibianus is comparable to a modern map: it attempts uniformity of scale over its whole area and, indeed, attempts it with some degree of success. Although individual features are shown in perspective (each to a separate horizon) the map on which they are placed is drawn in true plan: its scale is the same along one axis as along the other, the same along the top as along the bottom. The spread of the idea of drawing to a fixed scale revolutionized topographical mapping in the sixteenth century. It led to the marriage of two separate existing traditions, mapping and surveying. In the middle ages surveyors and map-makers had each gone their way, independent of the other. Surveyors measured lands, defined boundaries and drew up elaborate written descriptions of estates without presenting their information in map form. Map-makers drew maps without attempting to relate distances on the map to distances on the ground in any fixed proportions. Their maps were mostly pictorial; but they can be clearly distinguished from non-cartographic pictures of landscape in that they were pictures of a peculiar sort, showing landscape in plan or near-plan, viewed as though from a height in a position unattainable in the days before man's achievement of flight or aerial ascent. In the course of the sixteenth century the picture-map of this sort came to be fitted into the exactly surveyed framework of uniform scale provided by the surveyor. Now, as we shall see, it was surveyors who became the cartographers; hitherto it had been artists. Now it became possible to distinguish a topographical map from a bird's-eye view, and the two forms of landscape portrayal, which had the same origin, have followed different paths ever since.

This change from the picture-map to the scale-map based on measured survey was a development of great importance in topographical mapping, marking the change from the second to the third of its three main phases. The change was not something that happened only in sixteenth-century Europe: as we shall see, it had occurred in other societies long before but had failed to survive in a continuing tradition. Likewise the change from the first to the second of the three phases has occurred not just once but probably many times in the history of mankind. This was the change that brought pictures into map-making: universally recognizable representations of the features of the landscape replaced arrangements of signs or symbols which could be understood only by the initiated. Curiously enough, because our own topographical maps have moved away from pictures to a system of conventional signs that are often non-pictorial, these primitive maps, on which everything is shown by abstract symbols, may seem to us more familiar, more recognizable as maps, even more sophisticated, than the pictorial maps of the second phase. But this is to judge from superficial appearances that are a mere accident of historical development; in concept the symbol-maps lie far further from the topographical maps of today.

Overleaf

2, 3 Air photographs of contrasting portions of the area shown on plate I: the villages of Westbere and Hersden and the city of Canterbury. As these were taken in 1967 some features differ from the map of 1959, such as the new road on the west side of Canterbury.

Colour plates I–V

I Canterbury with its north-eastern surrounds from the Ordnance Survey map of Great Britain, scale 1:63360, published 1959.

II Zermatt and the Matterhorn from the official map of Switzerland, scale 1:50000, published 1944.

III Jerusalem as shown on the mosaic map of AD 560–65 at Madaba in Jordan. The complete map covers all of Palestine and some adjacent lands; east is at the top. In the city the main street from north to south is lined with arcades, the two sides being shown facing each other. Most of the streets and buildings can be identified.

IV Verona, Lake Garda and the Adige valley: part of a picture-map of Verona and its district, mid-15th century. Fortifications and roads are prominent; the detailed plan of Verona and the accurate mountain outlines are of particular interest.

V Pages from a manuscript atlas of Kiangsi province in south-east China, 18th century. The area shown is a single prefecture, the large town in the foreground being its administrative centre. Many Chinese local maps of this sort survive from the 17th century on; all are picture-maps.

I

II

III

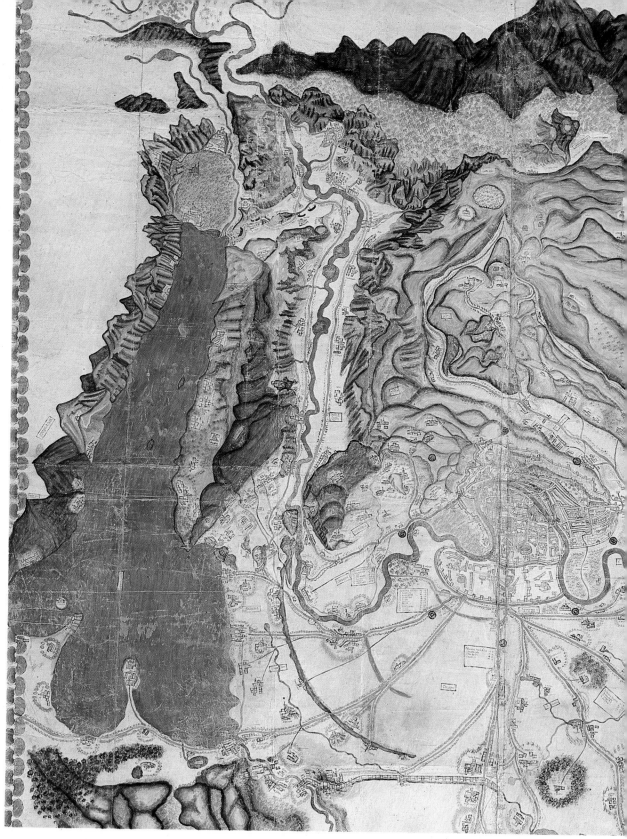

IV

九江府治圖

五老峰

香爐峰

三廬山

笑別寨

江寺

湖翠寺

大林寨

圓通寺

東林寺　西林寺

西

湖寺界廣湖至南西

仙壇　煙樹祠

棲賢亭

圓白祠　園亭

龍池寺

儒德亭

德化縣　濂溪亭

演武亭

羅星亭　曉池　德陽橋

鎮府　布政

九江府

昆世亭

九江關

江頂

人江口

蕩洪山

德安縣

義寧山

白蓮山界贛德至北西

水昌縣　寧亭

新昌縣　河治縣

石溪山

白沙洲

小石

龍坪市

V

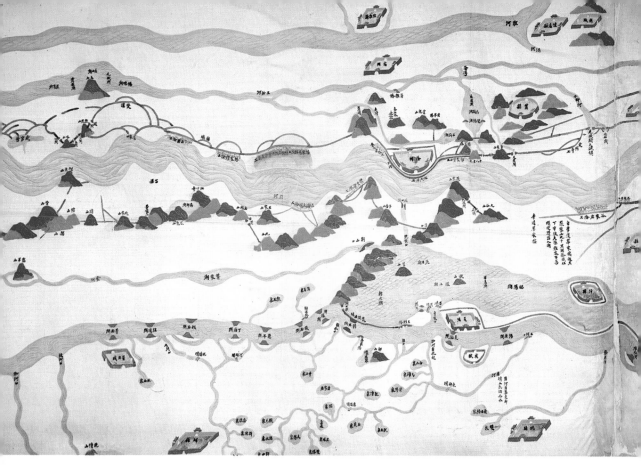

VI

VII

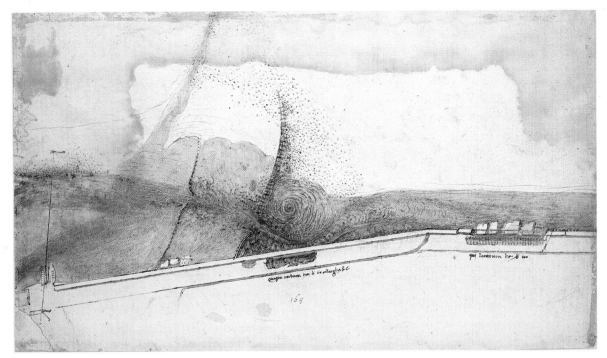

VIII

IX

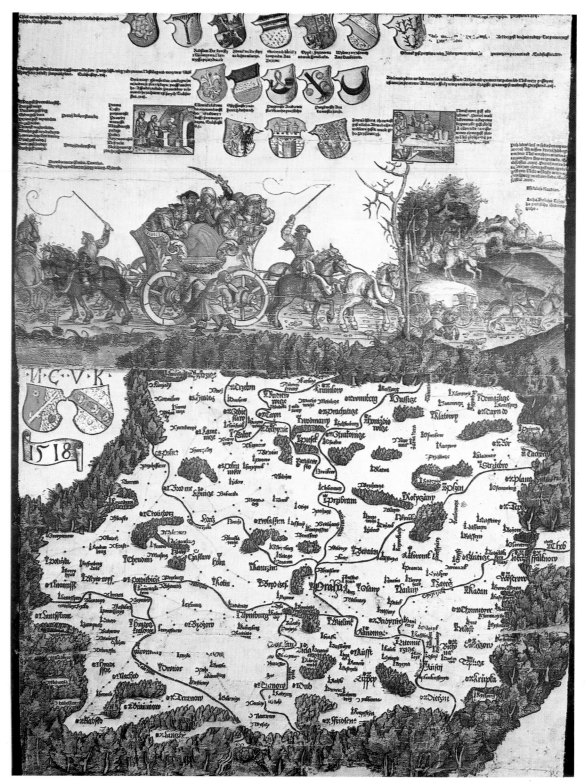

X

Colour plates VI–X

VI Part of the course of the Grand Canal which links Peking with Hangchow: this 18th-century Chinese picture-map shows the whole thousand-mile length of the canal on a long roll. Towns are shown by their walls and gates in bird's-eye view, a convention common on picture-maps from Europe as well as China.

VII Picture-map of the valley of Tepetlaoztoc in Mexico, 1583. Though drawn long after the Spanish Conquest this reflects Aztec traditions of cartography. The humped designs that dominate the map are ranges of hills; the symbols in the shield-shaped panels represent place-names.

VIII Plan by Leonardo da Vinci of part of a scheme to regulate the course of the River Arno, 1502–3. It demonstrates Da Vinci's interest in the movement of water, both as artist and as natural philosopher; also, in its particular use of mixed plan and perspective, his interest in the various cartographic techniques of his time.

IX Picture-map of Dover, probably drawn to illustrate harbour works about 1543. Inscriptions on the two inner harbours read 'And this parte of the harbour is both clensed and deped vij foot' and 'This herbour is enlarged and deped'.

X Map of Bohemia by Nicholas Claudianus, 1518; it was hand-coloured, probably at the time. South is at the top. The map is at the foot of a broadsheet containing also allegorical pictures, heraldic shields and notes.

Our story then is not one of a single developing tradition. It is one of changing methods of portraying landscape that seem to correspond to distinct stages in cultural development. When we speak of the progression from symbols to pictures and from pictures to surveys we are not thinking of a straight chronological sequence except in the context of a single society or culture. In the chapters that follow we shall be looking at each of the three phases of topographical mapping in turn, and we shall find ourselves moving from one age to another quite as rapidly as we move between different countries and cultures. It is an odd (though explicable) fact that some of the oldest maps to be discussed in the book are among the most advanced, belonging to the third phase of development, while most of those discussed from the first, primitive, phase are relatively recent productions.

We should thus see this progression – symbol, picture, survey – as corresponding to a growth of sophistication in man's response to the landscape around him. Dr Denis Wood has recently provided an interesting illustration of this by examining the way hills are drawn by young Americans today. He collected drawings of hills from three hundred people in North Carolina, aged between four and thirty, classified them according to the type of representation and analysed the results by age-groups. What he found is seen in ill. 4. The three main divisions he used to classify the drawings he described as elevation (simple outline as viewed from the side), oblique (similar outline but with shading suggesting perspective) or plan (represented as seen from above), progressing from the simple to the more elaborate in each division, a total in all of thirty-one types of representation. This same sequence of types could be divided slightly differently to match the three phases of topographical mapping that we have identified here: nos 1 and 2 (perhaps nos 3 and 4 as well) can be seen as symbols, the succeeding types down to no. 22 as pictures, and nos 23 to 31 as forms appropriate to maps based on surveys which would show the ground-plan of the hill. Divided in this way the analysis by age-group is even more striking than in Dr Wood's division: the 'symbol' forms were produced only by the youngest age-groups (aged four to seven), the 'surveyed'

4 Analysis by Denis Wood of the way children and adults draw hills. On the left are the various basic shapes produced by the 300 people tested. Only the oldest groups drew hills as if seen from above; only the youngest produced shapes that are not based on the actual appearance of hills.

forms only by the oldest groups (aged fourteen and above), while the 'picture' forms were produced by all ages and were the only ones to come from the middle age-group. Dr Wood's aim was to provide a basis for comparison with the historical development of topographical mapping; the slight shift in emphasis proposed here gives his results further interest and significance.

And indeed, although there are many episodes in our story that are obscure, and although future research has undoubtedly much to tell us that will modify or entirely change parts of it, this basic pattern of progression – symbols, pictures, surveys – seems clear beyond serious doubt. With this in mind let us start at the beginning and look at topographical maps in their most primitive form: the symbol-maps.

26

Part 1: Symbols

1

The beginnings

WHEN WE consider cartography in its most primitive form we have to think simply of designs or groups of objects representing features of landscape and arranged in a way that reflects the arrangement of these features on the ground. The simplest map need be no more than knotted twine or a string of beads that might serve as a mnemonic to remind a traveller of landmarks along the path of a journey. The maps made by the Bindibu aborigines of Western Australia are scarcely more elaborate. On the backs of their wooden spear-throwers are deeply carved series of patterns: semi-circles, spirals and concentric circles with connecting lines which in stylized form represent the waters in the area of the tribe's wide hunting-ground. In the western Pacific, on the Mortlock group of the Middle Caroline Islands, similar maps would take a more intimate form: the inhabitants would have their bodies tattooed with lines and other signs representing separate islands and island groups.

But of such highly schematic primitive maps what are probably the most remarkable, and certainly the most fully studied, are the stick-charts that come from the Marshall Islands, the eastern neighbours of the Caroline Islands. These consist of flat, open frameworks, more or less elaborate, made of the thin, flexible strips that form the centre ribs of palm-leaves; these are tied to make either straight or curved lines, and most have small shells attached either at intersections or elsewhere along the strips. These stick-charts are fragile and quite large: few are smaller than 2 feet (60 centimetres) square, and some are as much as 5 feet (1·5 5

5 Stick-chart from the Marshall Islands. Shells marking islands are tied to sticks that represent currents and lines of swell. These charts were made for instruction, not as actual navigational aids. This example measures 28 by 11 inches (70 by 28 cm).

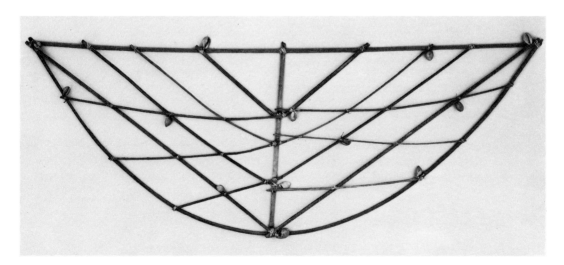

metres) long. Their principal purpose is to illustrate the behaviour of deep-sea swells; these change their form and direction as they approach islands and are thus an important guide in navigating the outrigger canoes used for voyages. There are three types of stick-chart: the *mattang*, which demonstrate the general principles involved, the *meddo*, which are maps of a small portion of the archipelago, mostly showing at least four islands, and the *rebbelith*, which are maps either of one or both of the two chains of the Marshall Islands (Ratak to the east, Ralik to the west) or else of the northern or southern half of both. The shells mark islands or island groups; the sticks represent either the lines of swell (these are usually curved strips), currents (usually short strips) or the line of the canoe's course. An example of a *meddo* is shown. The existence of these stick-charts was first reported in 1862 and they have continued in use alongside the navigational aids brought to the Islands by Europeans. In fact they perform quite a different function. They are not sea-charts to be used in plotting positions or setting a course; indeed, they are not meant to be taken to sea at all. It was once thought that at least the *meddo* were used in this way, but this was when their interpretation and use were mostly kept closely secret by the master navigators in each area. Rather they are for demonstration and instruction, providing a pattern, learned on land, that could easily be held in the head while at sea. But it is not so clear how widespread the use of the stick-charts was in the past; all known examples come from the southern half of the two chains of Marshall Islands, and reports of similar charts from the northern Marshall Islands, from the Caroline Islands and from Fiji have not been confirmed.

A map of a very different sort comes from Mer, one of the Murray Islands in the Torres Strait, between Australia and New Guinea. In former times boys would undergo an initiation ceremony at a spot where fourteen small stones were set in the ground. As part of the initiation they would be told the story of the legendary hero Malu, and of his voyage from one to another of the islands in the group until he came to Mer. This story was linked to the fourteen stones: each represented an island, and they were said to have been set in place by Malu

himself. The positions of the stones reflected the relative positions of the actual islands, but the overall layout can have corresponded only very roughly to their geographical pattern; the layout was in any case a little uncertain as by the time the stones were seen and recorded by A. C. Haddon in the late nineteenth century the ceremony had gone out of use and there was doubt over which islands some of the stones represented. Also on Mer was an oracle consisting of a group of stones, each with a large shell set on it; these represented the island's villages, and predictions about a particular village would be connected with the appropriate shell. One stone and shell set apart from the rest stood for the island as a whole, and anything that happened to them would affect the entire island. Unlike the stones of Malu's voyage the arrangement of the stones with shells bore no relation to the island's actual geography and thus were in no sense a map; but we should bear them in mind when we come to consider what purposes maps may have served in primitive societies.

So far we have looked at four examples of primitive maps, all drawn from the same quarter of the earth – Western Australia, the Caroline Islands, the Marshall Islands, the Torres Strait. They show something of the range of types of map that are possible when representation is entirely symbolic. They serve also to introduce some of the problems that confront us in maps of this sort. One basic difficulty is to discover how widespread the use of maps was – either maps of a particular kind or maps in general. Each of our four examples comes from a very restricted area and represents a particular local or tribal tradition of map-making. Even within the area of such a tradition, knowledge of the maps may be limited to certain people, who may be not at all willing to expound these mysteries to visiting seamen, missionaries or anthropologists. Equally the Europeans who happened to learn something of primitive maps need not have thought the information worth recording. Our knowledge comes mostly from the earliest explorers and colonial settlers in an area and depends very much on chance. This problem is one we shall return to; certainly it would now be difficult, probably impossible, to discover how far this kind of cartography existed among the primitive peoples of the world, or to discern any pattern in its distribution.

Nor is it easy to discover the origin and purpose of these maps. It would be altogether naïve to suppose that the only – or even the principal – reason why primitive people drew maps was to help them to find their way. It is perfectly possible that the Bindibu carved their spear-throwers with maps of the tribal waters as a handy *aide-mémoire* that one would always have with one, like the maps that were printed on soldiers' handkerchieves in the Second World War. But it is no less possible that the maps were symbolic in purpose as well as in form, representing an assertion of rights, a statement of sovereignty, rather as the English Commonwealth in 1649, having abolished the monarchy, put a map of the British Isles on the Great Seal in place of the picture of the king on his throne. And we must remember that the range of purely practical purposes that a map might serve in such societies would be rather wider than in our own, which takes a more limited view of causation: designs or objects representing houses, villages, fields, rivers and other features of the landscape would be an obvious vehicle for sympathetic magic, like the shells that represented villages on the island of Mer. Nor need what finished as a map have started as a map. The tattooed maps of the Caroline Islands could well have come into being simply by using the names of localities in the area (perhaps from an apparent resemblance in form, perhaps not) to identify parts of a traditional tattoo design that was really an abstract pattern or something else quite different. Because the stones on Mer were used as a map in recounting the voyage of Malu it does not follow that they were placed there for that reason, whether by Malu himself or by anyone else; they may originally have served some other purpose, being incorporated into the initiation ceremony at a later date. The true origins of cartography are scarcely ever revealed to us.

We must realize too that we can learn practically nothing of maps of this sort that may have existed in distant ages. Three of our four examples were executed in non-durable materials: wood, palm-fibres and the human body. And not one of the four could possibly have been recognized as a map without some explanation from the people who made it or used it. It may well be that some – perhaps a high proportion – of the many different sorts of

6 Bronze Age cup-and-ring markings at Drumtroddan Farm, near Port William, Galloway. Some prehistoric carvings may well represent landscape: most primitive maps are unrecognizable without explanation from those who made them.

unexplained rock carvings and groupings of stones that survive from prehistoric periods were really produced as maps. Promising candidates would be the rock-carvings in the so-called cup-and-ring patterns which date from the Bronze Age and are found throughout the British Isles and in the north-west of the Iberian Peninsula. 6 But it is most unlikely that we shall ever know; certainly no attempt to correlate elements of such designs with the neighbouring landscapes could carry conviction unless based on a rigorous statistical method, for in such cases the eye of faith will see what it wishes to find. Those who could have interpreted any maps that may exist among the relics of prehistory all perished long ago.

In fact the maps by primitive peoples for which we have the most plentiful and most widespread evidence are the most ephemeral maps of all, scratched in sand or drawn on some other surface as the clearest way of giving directions for a journey or of showing the plan of a district. Maps like this are mentioned in accounts of early explorers in many parts of the world: south-east Australia, the Congo, the Kurile Islands and so on. Maori warriors in New Zealand would describe past battles by drawing plans on the ground. Inhabitants of the Palau Islands, east of the Philippines, gave Spanish missionaries in 1696 an account of the area in cartographic form: some well-informed person from each district placed on a table as many stones as there were islands there, giving the name and size of each one and saying how far it lay from the others. Sometimes the symbols used in such maps suggest a real tradition of map-making in the area. Maps drawn in sand on the Angola coast would show the rising or setting sun to point direction. Karl von den Steinen, on an expedition to the Xingú River in central Brazil in 1884, had a map of the river system drawn for him in the sand by an old man of the Suyá tribe; cross-strokes marked the number of villages on each tributary. Partly because of what he learned from this map he made another expedition a few years later and

this time members of the Culiseu tribe drew him maps on which cross-strokes on rivers marked separate tribes or the speed of the current, circles marked houses, and rings of circles villages, corresponding to the arrangement of the round houses around a central space.

Traditions of symbol-maps seem to have been very widespread among the North American Indians. Stone-carved maps are known, but most were less permanent, drawn on deerskin or birchbark or scratched in the sand. They are reported from all parts of the continent. Colonel Henry Norwood tells how he saw such a map made in Virginia in 1649–50 by the 'King of Kickotank', one of a branch of the Powhatan Indians on Eastern Shore:

He took up a stick, with which he made divers circles by the fireside, and then holding up his finger to procure my attention, he gave to every a hole a name; and it was not hard to conceive that the several holes were to supply the place of a sea-chart, shewing the situation of all the most noted *Indian* territories that lay to the southward of Kickotank.

Henry Y. Hind, who was in Labrador in the mid-nineteenth century, describes a birchbark map drawn by a Naskaupi Indian and placed in a split branch beside a lake to serve as a guide to future travellers: arrows showed the route that had been taken, crosses marked camping-places and a large cross the site for winter quarters. The Blackfoot Indians of the upper Missouri scratched maps on the ground, basically showing rivers but with a fairly elaborate system of other symbols: yellow pebbles for day camping-places, black pebbles for night-camps, red sticks for enemies, black sticks with a scrap of tobacco (representing the pipe of peace) for friends, series of V-shaped marks for routes over land. In 1849 Chippewa Indians from Wisconsin sent a petition to the President of the United States accompanied by a birchbark map which showed the lake and rivers where they lived, each family being marked by a drawing of the animal, bird or fish that was its totem.

7

7 Reproduction of a map of the upper Xingú River in Brazil; this was drawn in sand by one of the Suyá tribe for the German explorer Karl von den Steinen, 1884. The names were spoken, not written.

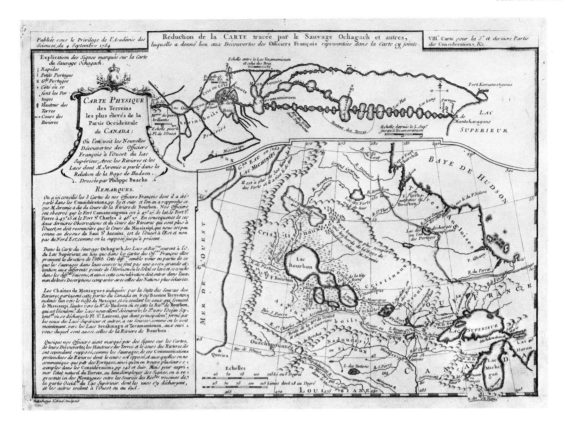

We have in fact direct evidence of quite a lot of map-making among the North American Indians, and hints of much more in maps made by European explorers of the seventeenth and eighteenth centuries that explicitly made use of information from Indian sources, much of it probably supplied in cartographic form. A good example is the far from accurate map of western Canada by Philippe Buache that was published in Paris in 1754: at its top is a map of the lakes and rivers between Lake Superior and Lake Winnipeg prepared from sketches made some thirty-five years earlier by a Cree Indian named (in French spelling) Auchagach and others. But the difficulty in using this as evidence of native cartography is obvious: it derives from maps made for the French explorers by Indians who were accustomed to contact with Europeans and may well have acquired the manner and style of their map-making from them. The same difficulty arises with most other maps we know of that were drawn by North American Indians: we can be fairly sure (though not certain even here) that the map scratched out for Colonel

8   Map of western Canada by Philippe Buache, 1754, partly based on the map of the Cree Indian Auchagach reproduced along the top. Early explorers refer to the North American Indians' use of maps, but this example must owe much to European guidance or interpretation.

Norwood in 1649–50 owed nothing to European example, but obviously we cannot say the same of the map of the Wisconsin Chippewas in 1849. This in fact is another very basic problem in the maps drawn by primitive peoples: in the vast majority of those recorded we cannot entirely rule out the possibility of outside influence in some form. It has been thought that even so idiosyncratic a production as the stick-charts of the Marshall Islands may owe something to European contact and European charts; at the very least European sea-charts may have influenced the development of the *rebbelith* type of stick-chart. Nor is it necessarily a matter of exposure to European culture alone. It is told how in the last century a smith of the Fula

33

people, living at Gadames in the north Sahara, showed the relative positions of settlements right across the desert – Tripoli, Gadames, Gat, Agadès, Timbuctoo – by drawing a map on which he marked meridians; here Arab influence may well have been responsible.

Certainly the more spectacular feats of cartography by primitive people seem to have followed contacts which could have introduced them to more advanced forms of mapping. George Forster, who assisted his father as a naturalist on Captain Cook's second voyage (1772–5), tells how a Tahitian named Tupaia drew a map covering much of the south-west Pacific; but earlier contact with the British is not impossible. Much more recently, the true size and shape of the Belcher Islands in Hudson Bay were discovered only in 1912–16, when they were explored by R. J. Flaherty; the reason for his expedition was an outline map of the islands made about 1895 by an Eskimo, Wetalltok,

which showed them in a form completely different from current charts, but which proved to be extraordinarily accurate. This says a great deal for Wetalltok's sense of topography as well as for his skill in mapping, but it tells us nothing of native Eskimo cartography, for he had been in close touch with Europeans and, probably, their maps – indeed, his map was drawn in pencil on the back of a missionary print.

9–11 Below are charts of part of Hudson Bay before (left) and after (right) the exploration of 1912–16 that established the true size and shape of the Belcher Islands. Above right is the pencil drawing by the Eskimo Wetalltok that initiated the expedition: a remarkably accurate map of the islands' oddly contorted outline. On the later chart we see a bay in the islands named after Wetalltok.

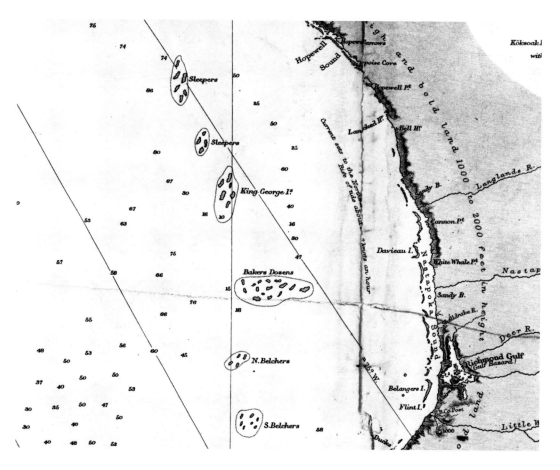

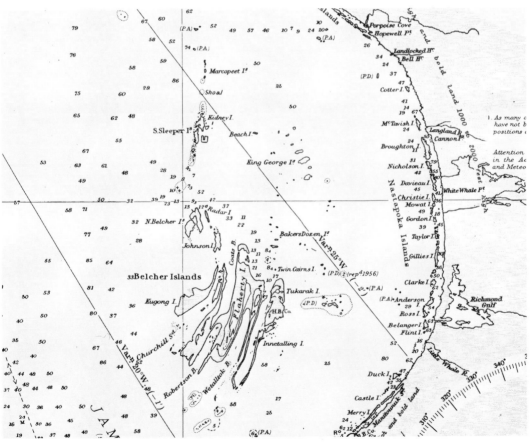

It need not be just a question of incoming Europeans introducing native peoples to more elaborate or more advanced forms of mapping than they used already. They may sometimes have introduced the very idea of a map. We should not assume that representing landscape by symbols, or by a few lines scratched in the sand, is so simple an idea that we can take an understanding of it for granted. Indeed, as we shall see in looking at picture-maps, even the simplest maps embody concepts that are far from universally understood, and some quite sophisticated societies had little or no knowledge of topographical maps of any sort. At the same time there is no room for doubt that at least some primitive societies were aware of maps and had cartographic traditions of their own before the arrival of Europeans. It is clear too that certain peoples on being shown European maps at once understood what they were and how

they worked; the Eskimos provide particularly striking instances of this. It has been suggested that the idea of a map might come most easily to peoples who were hunters, as they would necessarily have a close knowledge of topography over a wide area. In our present state of knowledge we cannot begin to draw conclusions about the distribution of map-making among primitive peoples or its significance, though it is worth remarking that a quite disproportionate amount of the evidence we have for maps among primitive peoples comes from (i) the North American Indians, (ii) the Eskimos, from Siberia to Greenland, and (iii) the peoples of Australasia and the south-west Pacific. But we have far more to learn about the map-making of primitive peoples. The present chapter does no more than scratch the surface, while pointing to some of the problems and difficulties that hinder our understanding.

2

From symbol to picture

WHEN WE look at symbol-maps to find the first traces of the picture-map it is difficult to know where to draw the line. The choice of the objects or the patterns used in even the most rudimentary symbol-maps can so easily have owed something to the actual appearance of the features they represent. We might say that there is a pictorial element, however dimly perceived, in the use of protruding objects to mark islands on some symbol-maps, such as the shells on the Marshall Islands stick-charts or the stones of Malu on the island of Mer. The introduction of outlines corresponding, however roughly, to the outlines of features on the ground, like the rivers and shore-lines on American Indian maps, is a distinct step in the direction of the picture-map and indeed it might well be seen as a particularly significant step in the development of cartographical concepts. And when we find maps executed in relief we can say that a genuine pictorial element has really started to emerge.

The map in relief might be very simple. There is an account of Indians on Vancouver Island having a plan of an enemy village drawn on the ground, the houses being marked with little heaps of sand; those of its chief and chief warrior were particularly clearly marked, as being the most dangerous. This is perhaps no nearer a three-dimensional model than the maps that use shells or stones to mark islands; but it is only a step away from the genuine, if primitive, maps in relief made by the Eskimos of Kotzebue Sound in Alaska in the early nineteenth century. Here the coastline would be drawn in sand with a stick and marked off in day's journeys; sand and stones were then used to build up islands and mountain ranges, taking their size and shape into account. Villages and fishing-places, on the other hand, were shown by sticks,

simply a symbol imposed on the three-dimensional model. Similar maps in relief and models built up from plans drawn on the ground are reported from the Tewa group of Pueblo Indians in Arizona and, outside North America, among the Maoris of New Zealand, who used stones to mark mountains, and the Tuaregs of the Sahara, whose maps were built up entirely in sand.

All these maps were ephemeral, made for a single occasion and not meant to last. Permanent maps in relief can also be found at an early stage in the development of map-making. The most striking case of this is the tradition of wooden maps among the Eskimos of Greenland. These are pieces of wood carved in the shape of coastal   12, 13 outlines. Some take the form of a bas-relief, carved on a flat background. Others are cut out along one edge of a stick, continuing along the opposite edge to give the finished map an extraordinarily contorted appearance. A complete map may consist of two sticks, one showing the mainland coastline, the other the offshore islands. Places where kayaks could be beached, or carried overland from one fjord to the next, may be marked; areas most familiar to the map-maker tend to be on a larger scale than the rest. They are examples *par excellence* of maps that could not possibly be used or even recognized as maps without explanation from the people who made them.

Maps oddly similar to the Eskimo carvings come from an utterly different source: Greek coins of the fourth century BC bearing a map in relief of the area behind Ephesus. We shall   14 return to these in the next chapter, but here they are a useful reminder that whereas the introduction of a third dimension into a primitive symbol-map may be seen as a first step towards pictorial representation, it does not

12, 13 Maps carved on wood by Eskimos of east Greenland. The two at the top form a single map: the shorter piece is carved in the shape of a stretch of coastline near Sermiligak, starting on one side and continuing on the other, while the large piece represents the islands lying offshore. Below is one of another sort, carved in bas-relief.

follow that a map in relief necessarily belongs to an early stage in cartographic development. Maps with this simple pictorial element of three-dimensional relief could well be made long after traditions of pictorial or even surveyed maps had been established, as we see from the moulded-plastic relief maps that we can buy today. It is with this in mind that we should consider the evidence from some more advanced cultures for maps made in relief. Here they may have represented early attempts at picture-maps of a rudimentary sort (whether or not preceded by a tradition of symbol-maps) or alternatively they may have been merely a part of a well established tradition of pictorial mapping. In India, where there was probably, at least in some areas, a tradition of pictorial mapping before the coming of the Europeans, a map modelled in relief on level ground is mentioned in the Skanda Purāna, one of a class of Brahmanical histories dating from the period of the European

14, 15 Long regarded as a meaningless pattern, the design on this tetradrachm from Ephesus in Asia Minor has now been recognized as a map of the region. Comparison with the relief map below shows how closely the features correspond. These coins were probably issued between 336 and 334 BC to pay the Persian army that occupied the area before the coming of Alexander the Great. No maps survive from ancient Greece apart from this and one other coin type, but the use of maps on coins argues general familiarity with the idea of cartography.

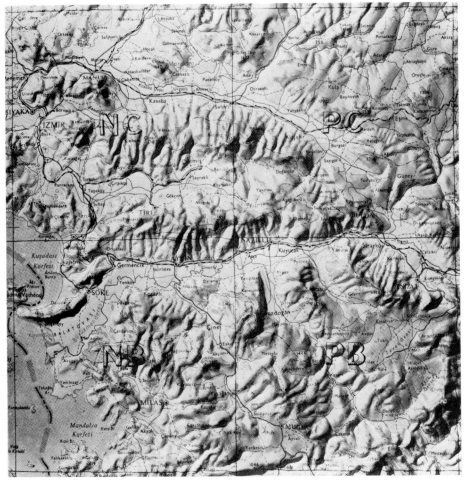

middle ages. Very much later, a three-dimensional map of the kingdom of Nepal made in the native tradition was presented to Warren Hastings, who was in India from 1750 to 1785. It was described in 1805 by Francis Wilford:

It was about four feet long, and two and a half broad, of paste board, and the mountains raised about an inch above the surface, with trees painted all round. The roads were represented by a red line, and the rivers with a blue one. The various ranges were very distinct, with the narrow passes through them: in short, it wanted but *a scale*. The valley of *Napál* was accurately delineated: but toward the borders of the map, every thing was crowded, and in confusion.

This was clearly a picture-map, and it may even have been influenced by European map-making.

We have accounts from Inca Peru of maps made in relief. One of the early Spanish historians of Peru tells how in the mid-fifteenth century Pachacuti Inca Yupanqui

ordered visitors to go through all the subdued provinces, with orders to measure and survey them, and to bring him models of the natural features in clay. This was done. The models and reports were brought before the Inca. He examined them and considered the mountainous fastnesses and the plains. He ordered the visitors to look well to what he would do. He then began to demolish the fastnesses and to have their inhabitants moved to plain country, and those of the plains were moved to mountainous regions, so far from each other, and each so far from their native country, that they could not return to it. Next the Inca ordered the visitors to go and do with the people what they had seen him do with the models. They went and did so.

Another historian, Garcilaso de la Vega, who published his work in 1619, gives a fuller description of relief maps that he had actually seen. The Incas, he tells us

understood very well how to paint and make models of each kingdom, and I have seen these models, with the towns and provinces depicted on them. I saw a model of Cuzco, with part of its province, and the four principal roads, made of clay and small stones and sticks. The model was according to scale, and showed the large and small squares, the streets, whether broad or narrow, the wards down to the most obscure houses, and the three streams which flow through the city. It was, indeed, a piece of work well worthy of admiration; as well as the model of the surrounding country, with its hills and valleys, ravines, and plateaux, rivers and streams with their windings, so well delineated that the best cosmographer in the world could not have done it better. They made this model that it might be seen by a person named Damian de la Bandera, who had a commission from the Royal Chancellery, to ascertain how many towns and Indians there were in the district of Cuzco.... The model which I saw was made at Muyna, a village called Molina by the Spaniards, five leagues south of the city of Cuzco.

But this, of course, must have been made long after the Spanish Conquest and may well have been influenced by Spanish map-making in its detail and its uniformity of scale. Whether the Incas had any maps – symbol-maps or picture-maps – other than those modelled in relief we simply do not know.

Nor are we much clearer about the role of maps in relief in the map-making of the Arabs. We have an interesting description of one in the writings of the fourteenth-century traveller Ibn Battúta who tells (from a contemporary's description) how Abú 'Inan, Sultan of Morocco from 1348 to 1358, was so concerned with the defences of Gibraltar, recaptured from the Christians in 1333, that

he gave orders for the construction of a model of it, on which he had represented models of its walls, towers, citadel, gates, arsenal, mosques, munition-stores, and corn-granaries, together with the shape of the Jebel itself and the adjacent Red Mound. This model was executed in the palace precincts; it was a marvellous likeness and a piece of fine craftsmanship.

We have a great deal still to learn about local and regional mapping in Islamic countries, but early Arab maps of larger areas suggest that any traditions there may have been are more likely to have been of symbol-maps than of picture-maps. Thus a medieval map of an area of central Asia around the River Oxus consists of a pattern of lines and coloured strips, circles and rectangles, with a hint of a pictorial element only in the use of blue strips and circles to mark rivers and lakes; towns appear as circles coloured red, green, yellow or purple, mountains by a flower-like shape. If the Arabs in the middle ages made symbol-maps rather than picture-maps this is of course not a reflection of their general level of culture, but rather of their particular tradition of abstract art: there would be no picture-maps because there were no pictures of any sort. In view of this we might see Abú 'Inan's model of Gibraltar as a striving towards pictorial representation within a tradition of symbol-maps. But this is to go far beyond our present evidence.

It would, however, accord with the far fuller evidence we have from China, as maps in relief there seem to date from the same period as the early development of the picture-map, and it may well be that modelling in a third dimension marked the transition from symbol to picture. The earliest reference is in a description of the tomb of an emperor of the Ch'in dynasty, Shih Huang Ti, who died in 210 BC: 'In the tomb-chamber the hundred water-courses, the Chiang [the Yangtze River] and the Ho [the Yellow River], together with the great sea, were all imitated by means of flowing mercury, and there were machines which made it flow and circulate'. Dr Joseph Needham makes the interesting suggestion that the models of mountains on pottery incense-burners and mortuary jars of the Han dynasty (206 BC – AD 220) may also derive from early maps in relief; and landscape models from the same period showing paddy fields and ponds have recently been discovered. Certainly by AD 32 maps for military purposes were being made with the mountains and valleys modelled in rice, a technique still known in the ninth century, when a treatise was written about it. There are also early references to maps carved in wood, though it need not follow that they were necessarily in relief; they may simply have had the features incised on the surface. We hear of wooden geographic maps of large areas in the fifth century – one that could be taken to pieces province by province, showing the whole of China when fully assembled – and again in the twelfth century when one that was certainly in relief was made by the historian Chu Hsi. And in a work written by Shen Kua in 1086 we have a very full description of the making of a three-dimensional topographical map:

When I went as a government official to inspect the frontier, I made for the first time a wooden map upon

16

16 Chinese incense-burner of the Han dynasty (206 BC–AD 220), its cover representing a mountain. An early example of a long Chinese tradition of modelling maps in relief.

which I represented the mountains, rivers and roads. After having explored personally the mountains and rivers of the region, I mixed sawdust with wheat-flour paste, modelling it to represent the configuration of the terrain upon a kind of wooden base. But afterwards when the weather grew cold, the sawdust and paste froze and was no longer usable, so I employed melted wax instead. . . . When I got back to my office in the capital I caused the relief map to be carved in wood, and then presented it to the emperor. The emperor invited all the high officials to come and see it, and later gave orders that similar wooden maps should be prepared by all prefects of frontier regions. These were sent up to the capital and conserved in the imperial archives.

From this varied range of maps in relief no clear pattern emerges. But we can say that three-

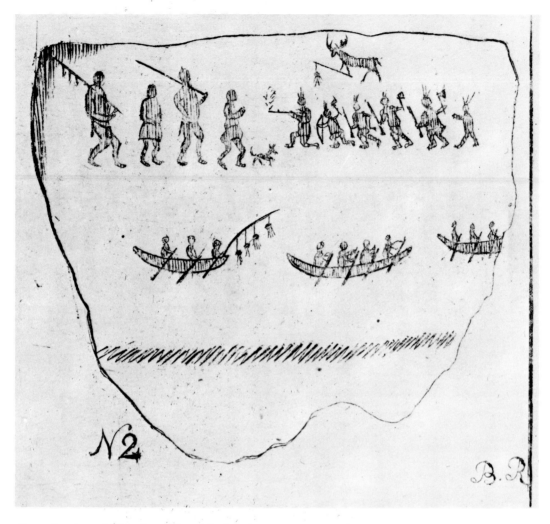

17 Picture-letter by Creek Indians of 18th-century Florida. It tells of an expedition by men of the Stag clan; one canoe carries home the four scalps of the defeated enemy, the stag at the top bears another, symbolizing the victory.

dimensional modelling is a pictorial element found even in very primitive forms of map; that more or less elaborate maps in relief appear certainly in China and perhaps also in India just at the point when the picture-map is starting to emerge; and that maps in relief may conceivably occupy an analogous position in the cartographic traditions of the Arabs and of Inca Peru. It is at least possible that the introduction of relief was one way in which the symbol-map developed into the picture-map. Another can be seen in the picture-letters or picture-stories of the North American Indians, though here again we can only point to possibilities, not demonstrate a clear course of development. We have seen how among the Indians of Labrador a birchbark map might be left in a split branch to guide future travellers; a birchbark letter or news-sheet might be left in the same way. Thus H. R. Schoolcraft, on an official expedition to discover the sources of the Mississippi in 1820, tells how, when passing through lands of the Ojibwa,

a small strip of birch bark, containing devices, was observed elevated on the top of a split sapling, some eight or ten feet high. . . . a symbolic record of the

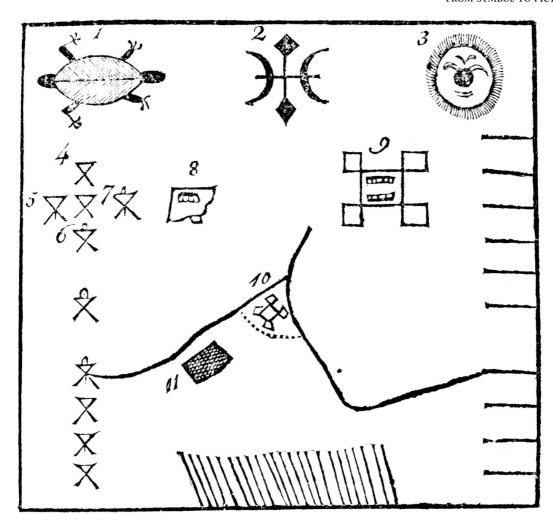

circumstances of our crossing this summit . . . so clearly drawn, according to their conventional rules, that the intelligence would be communicated thereby to any of their people who might chance to wander this way.

Messages of this sort were more or less conventionalized, but were basically pictorial. Two examples in differing styles, both dating from the late eighteenth century, are shown here. The first, from the Creek Indians of Florida, gives an account of an expedition by ten members of the Stag clan in three canoes: they ambushed a party of two men and two women (and a dog), killed them and went home with their scalps. Bernard Romans, who first published the picture in 1775, explains that 'the scalp in the

18 The deeds of a Delaware Indian chief told in a picture-story. The turtle, top left, identifies his clan; the symbols below show the number of scalps and prisoners taken in his expeditions, which included attacks on Fort Pitt and Detroit.

stag's foot implies the honour of the action to the whole family'. The second, found on a tree by the Muskingum River in Ohio, recounts the deeds of a chief of the Tortoise clan of the Delaware Indians: the number of his expeditions, his warriors, his defeated enemies, and his attacks on Fort Pitt and Detroit (both shown in plan). Neither is a map; yet neither is far from containing cartographic elements, and it is a very easy transition to a picture-map with a

17, 18

43

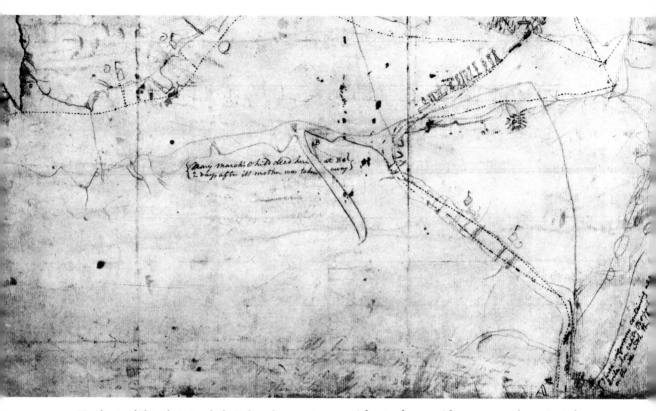

19 Shanawdithit, last Beothuk Indian known to white men, drew this map in 1829. It shows some of her tribe's wanderings along the rivers and lakes of central Newfoundland.

19 story, of the sort shown here. This is one of a series of maps drawn in 1829 by Shanawdithit, one of the Beothuk Indians of Newfoundland; she was the last of her race known to white men, for the Beothuks died out altogether not long after she was captured and taken to St John's. Her map shows part of the Exploits River in central Newfoundland; we see a party of Beothuks crossing the ice-covered lake with sleds and making for the wigwams on the farther shore. Shanawdithit's maps illustrate particularly well one of the difficulties we have already found in assessing primitive maps: the difficulty of distinguishing the native elements from the European influence to which the draughtsman may have been exposed.

We have far more to learn about the maps of the North American Indians, their types, distribution and chronology, and it is obviously risky to draw evidence, as we have just done, indiscriminately from the products of very varied peoples and cultures scattered over an entire continent. But these examples at least show that the distinction between the picture-story and the picture-map is not always easy to draw, and that the picture-story may well have been one way that pictorial elements were introduced into symbol-maps. Birchbark letters in map form are also recorded from the Yukagir people of north-east Siberia, and we shall see how in Aztec Mexico the picture-story seems to lead imperceptibly to the picture-map. And, of course, the map that is also a picture-story is not the product only of early forms of cartography. Later we shall be looking at a mid-fifteenth-century map from Italy that includes a picture of a wartime incident on Lake Garda, and in sixteenth-century Europe picture-maps illustrating military campaigns were quite common; examples among the illustrations to later chapters are the superb woodcut map of Lake Constance, about 1505, by the unidentified P.W., which shows scenes from the war of 1499, and a view of the siege of Enniskillen in 1594.

We can then only suggest that picture-stories and three-dimensional models may have been ways in which the symbol-map developed into the picture-map. Can we go further and point to traditions of mapping where we can see the change from symbol to picture actually taking place? It has already been said that maps in relief appear in China just as early picture-maps were developing there; in fact the earliest Chinese topographical maps known to us – the recently discovered maps from Chang-sha, dating from the second century BC – seem to lie on the borderline between symbol-map and picture-map and it could be argued, taking these with the evidence of maps in relief, that we can see the change from symbolic to pictorial representation taking place in Early Han dynasty China. But the Chang-sha maps contain pictorial elements – we have no pure symbol-maps from China – and they are best discussed later as part of the pictorial tradition in Chinese mapping. We have, however, in Europe a most interesting early local tradition of mapping which clearly shows symbols being replaced by pictures; this is in the rock-carvings of the Valcamonica.

The Valcamonica is a valley running up into the Alps from Bergamo in north Italy. It is remarkable for its ancient rock-carvings; about 20,000 are known, dating from the second and first millennia BC. Although similar carvings are found elsewhere in the Alps, those of the Valcamonica are unmatched both in their number and in the competence of their workmanship. There is great variety of subject – human figures, animals, carts, weapons and so on – and one common type consists of maps of fields and villages. The earliest of these are large designs of lines, dots and stippled surfaces, and they could hardly have been recognized for what they are were it not that later carvings in the same tradition show houses and other features pictorially, in simple elevation. We see examples of both forms of map, the symbolic and the pictorial, in ills 20 and 21, which show just how detailed and elaborate they are. Their purpose is lost to us; it might have been legal, as an assertion of proprietorship or lordship, it might have been religious, as a vehicle for sympathetic magic, it might have been something quite different. The change in style occurred about the middle of the first millennium BC, and it is particularly interesting as being simply one aspect of a change in the general artistic style of the carvings as a whole: representation became more realistic, and objects such as ploughs and carts were shown as if seen from the side instead of from above. We

20 Carving from the Valcamonica in north Italy, 2nd millennium BC. It is a map of fields, paths and houses, as appears by comparing it with the valley's later picture-maps (ill. 21).

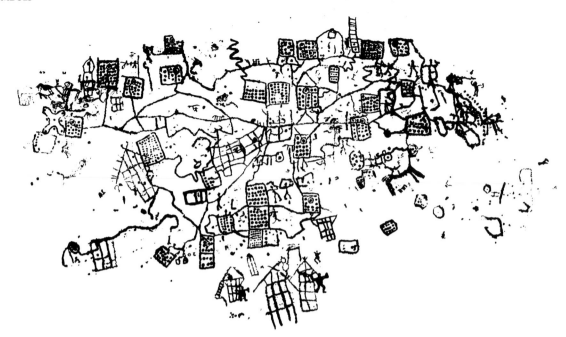

21 Picture-map of Bedolina in the Valcamonica, carved in the 1st millennium BC. Amid stippled square fields we see animals and houses, some with people in the upper floors reached by ladders.

can see how this corresponds to showing houses on the maps by pictures of their gable-ends instead of by square blocks. This general change in style may have come about through a simple advance in artistic methods and concepts among the carvers of the valley, or through new contacts and influences from outside. But otherwise the evidence of Valcamonica is very clear: we see the symbol-map giving way to the picture-map as part of a wider artistic or even cultural change.

Little systematic work has been done on the maps of primitive peoples. W. Dröber, one of the few writers to cover the subject as a whole, adopts a classification based not on the form of representation but on the physical material of the map: carved on stone, scratched or built up on the ground, made so as to be portable (such as the Marshall Islands stick-charts), drawn on skins or birchbark or, finally, drawn with pencil and paper or other materials provided by Europeans. This need not be wholly unrelated to the development from symbol to picture that is proposed here: the more adaptable and

flexible the material used, the more it lends itself to production of pictures instead of symbols. We have just seen from the Valcamonica that the picture-map may perfectly well emerge in a tradition of rock-carving, but it is likely that even here other less durable materials would also be used for artistic work, and the changes in the style of the carvings may well have originated in other media.

The proposed sequence of development from symbol to picture need not imply that traditions of picture-maps in particular societies always, or indeed often, took their origin in earlier traditions of symbol-maps. It is most unlikely, for instance, that symbol-maps lie behind the pictorial maps that survive from the medieval Netherlands and England; as we shall see, their antecedents – if indeed they had any at all – are to be sought rather in building-plans and in non-cartographical diagrams. What is suggested, however, is that when the most primitive peoples turned their hands to map-making the result would normally be a symbol-map, whereas a more advanced society would tend to represent features of landscape by pictures. The two forms of mapping correspond to distinct stages of cultural or artistic development. To this extent, and to this extent only, we can speak of topographical mapping as progressing from symbols to pictures.

Part 2: **Pictures**

3

The classical tradition

WHEN WE come to picture-maps we find as great a variety of form and style as in symbol-maps. But however varied they may seem to be superficially, the same cartographical concepts lie behind them all. First, no particular importance is attached to uniformity of scale. Often the notion is deliberately flouted as parts of the map considered especially important are drawn disproportionately large; the late Professor E. G. R. Taylor once shrewdly remarked that we might call such features 'grossly exaggerated', but the map-maker would probably say that he had 'carefully emphasised' them. And secondly, if the map consists of more than the merest outlines all features are shown by pictures. Sometimes, of course, there are no pictures on the map because it shows nothing but routes, boundaries or other ground-level features that appear simply as lines; this is the diagram-map. But if any three-dimensional features are included they will appear pictorially, not as symbols or in ground-plan. We see these productions as maps, not just as pictures, because they represent landscape as if seen from a viewpoint (or more than one viewpoint) unattainable in reality. At the same time all but the very simplest are essentially works of art, firmly rooted in the artistic tradition of the society that produced them. If picture-maps from sixteenth-century Mexico and fifth-century Rome look quite different this is not because they differ cartographically in any significant way, but because all Aztec pictures look different from Roman ones. In picture-maps we are dealing with the products of artists, artists who would probably have been unable to understand our distinction between those of their works that we call maps and the others that we do not, and who certainly used the same styles for both.

We know far more about picture-maps than we do about symbol-maps, but at the same time what we know is curiously patchy. This is partly because scholarly research on them has been limited to a few individual maps and closely defined groups; we have much more to learn about the picture-map, just as we have about the symbol-map. But it also reflects an odd patchiness in the origins of picture-maps, and this, as we shall see, may not be merely an accident of survival: some societies that we might have expected to produce picture-maps seem to have produced no topographical maps at all – the concepts behind this form of cartography or (put another way) this artistic technique simply eluded them. With this in mind it must be said at once that the title of this chapter ought to be followed by several question marks: it may be a complete misnomer. We have picture-maps from Mesopotamia in the third millennium BC; we have them from Egypt in the second millennium; from the heartlands of the Assyrian Empire in the seventh and sixth centuries BC; from outposts of Greek civilization in the fifth and fourth centuries; from the Roman Empire from the second to the sixth century AD; from Italy from the thirteenth to the fifteenth century. Apart from the last we cannot say beyond all doubt that there was a conscious tradition of picture-maps even within any one of these societies: the maps surviving may all be unrelated to each other. And we certainly cannot prove a connection between any of the groups. All that we can say is that it is possible that the idea of picture-maps may have passed from one to another and not quite beyond the bounds of possibility that a single tradition links them all, that an unbroken thread of pictorial mapping stretches over four thousand years or more, all the way to

Renaissance Italy from the empire of King Sargon of Akkad.

From Mesopotamia we have a substantial number of clay tablets and stones incised with maps and plans, dating from the third to the first millennium BC. Nearly all of them show nothing but field-boundaries or other ground-level features and outlines; by the second millennium some at least were being drawn to consistent scale from measured surveys, and we shall be looking at the group as a whole in this context in chapter 8. But it is interesting, and may be significant, that the earliest of all these incised Mesopotamian maps is the one that is most clearly a picture-map, if of a rudimentary sort. The chronology of early Mesopotamia is beset with problems, but this map is likely to date from about 2500 to 2300 BC and it is probably the world's oldest recognizable topographical map. It was found at Nuzi, in northern Mesopotamia, in excavations of 1930–31. Its pictorial

features are two ranges of hills and the rivers that they flank; the hills appear as double rows of dome-like marks, the rivers as series of parallel lines, a convention that also appears on a few of the later tablets. At least three towns are marked by circles and named, and a note in the middle of the map gives the size of an estate. The site of this property and the rest of the area covered by the map is not certain, but it was probably not far from Nuzi, and may have lain

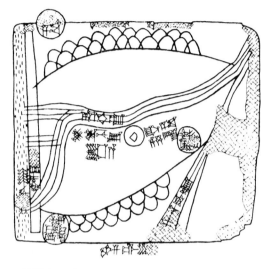

22, 23 The world's oldest known topographical map: a clay tablet from Nuzi in north-east Iraq, 3rd millennium BC. It is a relatively advanced picture-map, showing mountains as seen from the side, rivers by series of parallel lines. Its detail is clearer in the drawing (right).

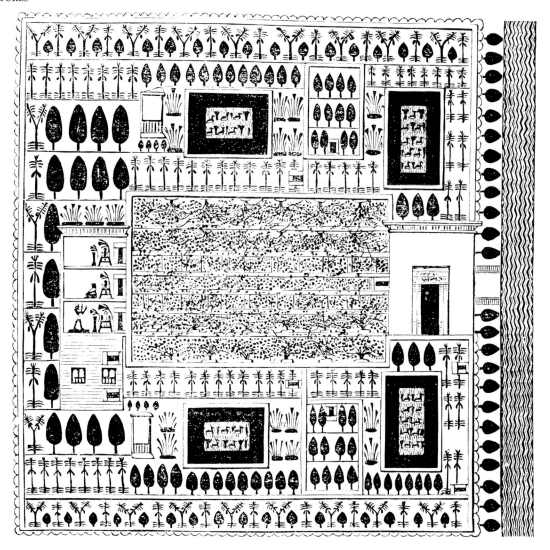

between the Zagros Mountains and the hills running through Kirkuk in north-east Iraq.

The next oldest picture-maps we know are Egyptian. Ironically, the earliest Egyptian map is one not of this world but of the next; it is a diagram-map without pictorial features and is painted on the bottom of each of nine coffins which date from about 2000 BC. It illustrates the text known as the 'Book of the Two Ways', the oldest known of the various Egyptian guides to the afterlife, and it has been suggested that it derives from maps of parts of the Nile; this may or may not be the case, but it certainly suggests that the Egyptians were familiar with picture-maps at a period a good deal earlier than their

24 Plan of a garden from Upper Egypt, 2nd millennium BC. At the centre is a vineyard with buildings on both sides, on the right the tree-lined bank of a river or lake. On four small pools are ducks and water plants.

earliest surviving maps of actual landscape. Of these by far the most thoroughly studied is one on a number of fragments of papyrus that are now in the Museo Egizio at Turin. Like the maps on the coffins it shows roads and stream-beds – one is marked with flecks that probably represent stones. Hills are shown by lines of peaks and houses also appear in elevation as rectangles with a little square at one side to represent a door. One of the inscriptions reads 'The hills from which the gold is brought are drawn red on the plan', and others refer to a hill of silver and gold and to roads leading to the sea. The map has been variously interpreted in the past, as showing the site of the tomb of the Pharaoh Seti I (who died in the late fourteenth century B C) or as a map of some of the mines in Nubia that were the Egyptian kingdom's principal source of gold. In fact it was drawn for a lawsuit about the portrait-statue of a king that was begun but, for some unexplained reason, not completed. The area it shows is Wadi Hammamat, halfway along the road between the royal capital at Thebes on the Nile and the Red Sea port of Kosseir. A surprising number of the features it shows can still be identified on the ground, among them a monument to Seti I and a well that appears as a black circle on the map. It probably dates from about 1200 B C.

Much less work has been done on other picture-maps from ancient Egypt, though they are of no less interest than the plan on the coffins and the Turin papyrus. They have never been investigated as a group, so it is not clear either what period they span or how many survive. However, we certainly know of a number of plans of buildings, grounds and gardens which come from Upper Egypt in the second half of the second millennium. In principle these consist of ground-plans which have superimposed on them pictures of vertical features. Thus buildings appear in elevation, with doorways and decorated columns; trees, shrubs and plants are also shown pictorially, as seen from the side, and the number of clearly different species is remarkable, as we see from 24 the plan of a waterside garden. Water, as here, is normally shown by series of parallel lines, either wavy or zigzag. There is some variety in the plans; thus a house can be shown by ground-plan alone, with only its porticoes, ornamental trees and the contents of its store-rooms appearing as pictures. But there are enough similarities in style between these plans to suggest a distinct tradition or fashion, though how widespread it was or how long it lasted we do not know. Many of the plans are drawn on the inner walls of tombs, but one from 25 Thebes, now in the Metropolitan Museum in New York, is drawn in red and black ink on a

25 Tablet from Thebes painted with the plan of a garden. In these plans from Upper Egypt the same conventions keep recurring: compare the water and trees with ill. 24. They point to a clear tradition of picture-maps in ancient Egypt, but whether it often extended beyond plans of gardens and houses we do not know.

wooden board surfaced with plaster. It may represent the entrance to a temple and it is of particular interest because although it is not drawn to a consistent scale it does give some of the dimensions of the features shown.

One picture-map from the walls of the palace at Thebes shows in the foreground an army drawn up and in the background the ground-plan of the double ditches or moats around a town; in the centre is the town itself, represented by a picture of its walls as seen from the side. In its combination of plan and elevation it follows the same principle as the plans of gardens and houses, but in its subject-matter it is reminiscent of a quite different artistic tradition that occasionally made use of the picture-map. This was the sculpture in bas-relief that was produced in the Assyrian and Persian empires in the seventh and sixth centuries B C. It had no

obvious connection with the ancient Mesopotamian tradition of incised maps on clay tablets
and stones, for these had long since lost any
pictorial element. And indeed the crudely
scratched picture-map from Nuzi two thousand
years earlier simply cannot be compared with
the bas-reliefs, which were works of artistic
vision and great craftsmanship. One from
Nineveh shows a triumphal procession on the
capture of Madaktu by Assurbanipal and it
26 gives us a very clear picture-map of the town;
we see its walls with houses inside and out, the
palm-trees around it, the small stream on the

26 Bas-relief from Nineveh, 6th century BC, showing
the triumphal procession on Assurbanipal's capture
of Madaktu, with warriors, harpists and (left) the
wheels of a chariot. Between is a picture-map of the
walled city with palm-groves and houses in its
suburbs.

farther side, and on the nearer side the main river
with fish visible in the water.

This last detail – the fish in the water – is one
that we find on picture-maps of very different
ages and cultures. It reminds us that although

the picture-map appears in a great variety of artistic traditions and styles it produced some techniques that recur in widely different contexts. Its particular way of portraying landscape created certain problems for the artist, and often the same methods were hit on to solve them. Thus on some of the Egyptian plans the design demanded that certain features should be drawn sideways – for instance a line of trees running away from the viewer could best be shown in this way, as we see in ills 24 and 25. In other words the plan would have to be turned different ways if parts of it were to be seen the right way up. This is a device that occurs time after time in picture-maps; we see it, for instance, in a twelfth-century plan of Jerusalem and in a fifteenth-century map of Inclesmoor in Yorkshire. We find a particular use of it in Assyria in the seventh century BC on a bas-relief of the deeds of Sennacherib. Here the problem was how to show the king and his soldiers marching along the bed of a stream between high, wooded banks; and it was solved by, as it were, opening up the picture along the line of the stream, showing both banks as they would be seen from the stream-bed, the farther one the right way up, the nearer one upside down. This method of showing on a picture-map features that would mask each other if they were shown in straightforward pictorial perspective also occurs over and over again – it represents a solution to what is indeed a recurrent problem in pictorial mapping. Thus we see it used to show both sides of city streets in the Madaba mosaic of the sixth century AD and again in the plan of Rome that was drawn about 1320.

From ancient Greece we have to assume a knowledge of topographical maps from very slight evidence: two coin designs are the only Greek maps to come down to us. The first is from the Greek colony of Zancle (now Messina) in Sicily, and dates from about 525 to 494 BC. The name Zancle came from a local word meaning sickle, and it refers to the curving sandbank that forms its harbour. On the coins we see the shape of a sickle blade enclosing a dolphin and the town's name. This was apparently meant as a map of the harbour, for one or two types of the coin have rectangular blocks on the sickle-shape that can be interpreted as buildings on the bank. The dolphin would be part of the picture-map, like the fish on the bas-relief of Madaktu. The other

Greek coin map is on a series of tetradrachms issued at Ephesus in Asia Minor, probably about 335 BC. It is only recently that their markings have been recognized for what they are: a map of the area behind Ephesus, some 90 square miles (23,000 hectares), showing the hill ranges in simple relief, sometimes with raised lines marking the rivers. One of the earliest and clearest types is illustrated, and we see at once its similarity to the area's pattern of relief. But although we know no more of topographical maps among the Greeks it is reasonable in this case to assume that they really did exist: to choose a map as the symbol of a city on a coin argues real familiarity with the idea of a map. That Greek maps normally took the form of picture-maps is very likely; but more than this we cannot say.

The same problem in a less extreme form confronts us when we turn to the Roman Empire. As we shall see in chapter 8 elaborate scale-maps based on measured surveys were being made at Rome and elsewhere from the first to the third century AD. Behind this there probably lay a long and continuing tradition of Roman picture-maps, but very few survive and most of them date from the fifth and sixth centuries. A mosaic of the mid-second century AD at Ostia shows a river, dividing to form a delta at its mouth, with upstream a pontoon bridge viewed obliquely from above; it can be seen as a simple picture-map, meant to represent perhaps the Nile, perhaps the Rhone. A more elaborate picture-map appeared in a painting of the early third century; the original is now lost, but an apparently accurate engraving was published in 1764. It is a bird's-eye view of a port, showing monumental buildings (some of them named) both in the town and on an island, and with a pier or mole on which there are arches, columns and statues. It probably shows Pozzuoli, near Naples. A manuscript of the works of Vergil, dated 420 and now in the Vatican Library, contains two picture-maps, one of Sicily, one of the coast of the Aegean showing five small islands and two cows. Both are crudely drawn and painted. Towns are represented by buildings and walls, a convention that we find also in the *Notitia dignitatum*. This is a treatise on imperial administration that we know only from late-medieval copies (at several removes) of a version

14, 15

35
52

III

36

compiled after 395 AD; it includes maps with stylized pictorial representations of towns and forts. Again, the Peutinger Table, which we shall look at in more detail along with other itinerary maps in chapter 9, is essentially a picture-map: mountains are shown by undulating lines of peaks, forests by rows of trees, and the conventional signs that mark hostelries and other staging-posts are pictorial, while at Ostia is a picture of the port based on its actual appearance. Like the *Notitia dignitatum*, no copy of the Peutinger Table survives from the Roman period, and we know it only from one made in the eleventh or twelfth century; it may originally have been compiled in the first century AD, being successively revised to reach the form reproduced in our copy about the middle of the fifth century. These examples vary a good deal in style and date, and they show at least that the picture-map was not unknown to the Romans. But more important as evidence of a strong tradition of pictorial mapping in the Roman Empire are the illustrations to the treatises on surveying and the Madaba mosaic.

In the Roman treatises on surveying we have again a group of texts that are much older than the earliest surviving copies. Here, however, the chronological gap is not so great: of the two earliest manuscripts with illustrations one dates from the sixth or seventh century, the other from the ninth. The date of the text itself is harder to establish, for it consists of a number of different treatises on surveying (put together in such a way that we cannot tell precisely how many) of which some date from the first and second centuries; but in the form known to us the work cannot have been compiled before the mid-fifth century and the illustrations may not have been drawn until then. Many of these illustrations are simple diagrams, setting out particular problems or methods of surveying. Others are straightforward pictures, such as those showing the different sorts of boundary-stone that were used. But some are indisputably picture-maps, giving the layout of fields and boundaries in plan, but showing rivers, roads, bridges, towns, mountains and other features pictorially. The maps are coloured: water is blue or blue-green, roads red, brown or sometimes green, buildings have red roofs, mountains are mauve or sometimes brown, or else green if they

are wooded. A very important point about the maps, and one not always appreciated, is that these are not the sort of map that surveyors themselves would be expected to draw. Roman surveyors did draw maps (*formae*) – indeed the treatises tell them how to do so – but the maps they drew were of a different and far more sophisticated sort: surveyed maps, drawn to consistent scale, which we shall discuss in chapter 8. The maps in the treatises were not models for the surveyor to follow; they merely served to illustrate particular points in the text, and they represent a technique and a tradition quite different from the surveyors' own maps. Professor O. A. W. Dilke, who has done so much to analyse and explain the illustrations in the treatises, has shown that picture-maps are used there particularly to set out legal definitions and problems of surveying and mapping near towns. Nine of the maps give names to the places they show, and three of these places can be identified. One is reproduced here; it shows the area around 'Colonia Axurnas', that is Terracina on the Italian coast halfway between Rome and Naples. On the left we see the rectangular pattern of fields laid out on the system of centuriation that is described in chapter 8; beyond them the Appian Way enters the town, represented by the picture of its walls alone, while behind the town are mountains and, in the top left corner, the edge of the Pontine Marshes ('paludes'). It may be that many or even most of the maps in these treatises derive from picture-maps of actual places, and that this is concealed either by the omission of the names that would identify them or else by distortions arising in successive copying. But in any case these illustrations point more convincingly than any other source to a real tradition of picture-maps in the Roman Empire: picture-maps were so familiar, so readily intelligible, that they were used in these treatises in preference to illustrations in the style of surveyed maps. The maps in the treatises are very significant in the history of the picture-map in Europe.

So too is the Madaba mosaic map. This was made probably between 560 and 565 for the floor of a church at Madaba, now in Jordan, some fifteen miles east of the Dead Sea. When it was made it was probably some 72 feet long and 22 wide (22 by 7 metres) and showed the whole of Palestine and adjacent lands from Damascus to

27 Roman surveyors made detailed scale-maps as part of their work. But Roman manuals on surveying were illustrated by picture-maps, some of identifiable places. 'Colonia Axurnas' in this example, a 9th-century copy of a late-Roman original, is Terracina, on the coast south of Rome.

Alexandria, taking as its centre the column that it shows just inside the north gate of Jerusalem. It has however suffered a great deal of damage over the centuries; we now have only one large irregular fragment some 16 by 34 feet overall (5 by 10½ metres), together with three small ones. The greatest loss has been in the area north of Jerusalem; the sections surviving cover much of Judaea, the River Jordan and the Dead Sea and, to the south, the mouths of the Nile. Although it covers a wide area it is essentially a topographical map of pictorial type. It has no fixed scale – in fact the scale is largest at the centre, smaller at the edges, something we often find on picture-maps. All detail is pictorial. Although mountains are shown in strongly outlined blocks of mixed colours that might be taken as ground-plans rather than elevations, the fact that the farther outline of a block tends to be humped or even jagged, the nearer one curved but smoother, suggests that representation in bird's-eye view was what the artist had in mind. Small towns and villages are shown by what are in effect conventional signs: two, three or more towers seen from the side and linked by lengths of wall. But ten cities are shown in bird's-eye view, the principal street of each being opened up so as to show the façade on each side. In the views of Jerusalem and Neapolis (Nablus) the principal buildings can be identified – indeed, Dr Avi-Yonah, in his detailed study of the map, has been able to identify all five of the streets and most of the thirty-six buildings with which it represents Jerusalem – and possibly all the bird's-eye views of cities are based on their actual appearance. Other detail is decorative: palm-trees, a stag-hunt, boats and fish in the Dead Sea. The whole was a large, colourful and elaborate work of art, skilfully composed; thus four different shades of red were used in it, five or six of blue-green and so on. Its purpose was clear: to illustrate biblical history, 'the instruction of the faithful' as Dr Avi-Yonah puts it. Thus the places it shows are either important cities, or the sites of events in the Bible or church history, or else have been put in to fill spaces otherwise blank. In addition the inscriptions (which are all in Greek) include biblical and historical notes. That such a map should have been made for such a purpose in a distant corner of the Empire again argues a general familiarity with the idea of the picture-map.

One particular point of interest about the Madaba mosaic is its possible connection with maps of the Holy Land drawn in late-medieval Europe. Several are known, dating from the twelfth century onwards, but how far they are related even to one another and to contemporary written accounts of the Holy Land has never been fully worked out, and would be a more complicated investigation than might appear at first sight. Two, very unlike in appearance, are by Matthew Paris, an Englishman who was a monk at St Albans in the mid-thirteenth century, a man of ingenious mind whose work we shall be referring to again later. One of his maps exists in a single copy which may be little more than a draft or sketch. It is unusual in being oriented to the north; most medieval maps of the Holy Land, like the Madaba mosaic, place east at the top, probably because Europeans normally approached Palestine from the west by sea. It marks a great many places known from the Bible or the crusades, and shows mountains and forests in simple picture form. It is far from accurate in either its proportions or its details – the River Nile, for instance, is named Tigris – but it is of great interest for both its form and its content. Matthew Paris's other map is less workmanlike but a good deal more picturesque. We have three versions of it that he made himself, all slightly different. But they all consist of little drawings and long inscriptions, and are dominated by the town of Acre, with its walls and some half-dozen of its principal buildings shown in elevation; this occupies about a sixth of the whole surface of each map, but the rest of

28

28 Map of the Holy Land, mid-13th century, by Matthew Paris, English monk, historian and artist. Acre, the chief town still in the Crusaders' hands, is shown as a vast walled enclosure; outside the walls are a camel for local colour and a cemetery shown as a circular enclosure with graves in it. Jerusalem, effectively lost to the Christians since 1187, appears as a much smaller walled square, top right; beyond it is Bethlehem, with star above, and along the top border are the River Jordan, the Dead Sea and Damascus, surrounded by trees. Some of the inscriptions are Latin, some French.

the Palestine coast is also shown, as well as the River Jordan, the Dead Sea and some inland towns – Damascus, Jerusalem, Bethlehem and Cairo among them. The relative positions of the places shown are very inaccurate, even allowing for the inconsistency of scale.

There is more likely to be a demonstrable connection between the Madaba mosaic and a map of the Holy Land that we know through two versions from medieval Italy. One was drawn at the end of the thirteenth century. The other was drawn by Pietro Vesconte as one of a series of maps illustrating the *Liber secretorum*

fidelium crucis by Marino Sanudo; we have several copies of this work, and in one manuscript it is dated 1320. Like the Madaba mosaic this map shows the division of Palestine between the twelve tribes, there seems to be at least an echo of the mosaic in the places that are marked and mountains are shown by shaded splodges that could be related to the outlined blocks of the mosaic. On the other hand the Italian map includes places made known only by the crusades, and some of its inscriptions are connected with the account of the Holy Land that was written by Burchardus de Monte Sion between 1271 and 1291. It would be of great interest to know how far the various medieval maps of the Holy Land are related to each other – whether, for instance, this Italian map and the maps of Matthew Paris owe anything to a common original – and if this was to show any clear connection with the Madaba mosaic this would provide the firmest link we have between the picture-maps of medieval Europe and those of classical antiquity.

It is not surprising that medieval maps of the Holy Land were produced in Italy, for Italy was by far the most map-conscious part of Europe in the middle ages. We see this in its geographic maps: portolan charts from the late thirteenth century onwards for Mediterranean navigation, the maps of Ptolemy's *Geographia*, which was first translated into Latin in 1406 by an Italian, Giacomo d'Angelo, maps of the whole of Italy which latterly drew on both these traditions, and others. But in addition there were topographical maps of places in Italy. These fall into three groups, all of them picture-maps. First are plans or bird's-eye views of towns; these evolved a very distinct tradition of their own, with possible connections elsewhere in Europe, and we shall look at them separately in the next chapter. Then there are district or regional maps of substantial areas of country, mostly centred on particular towns in north Italy. Both of these groups – the town views and the district maps – have been the subject of much research in Italy and elsewhere; they have been published, analysed and discussed, so that we have a fairly clear picture of their number, chronology and development. On the third group – maps and plans of particular localities and plots of land – practically no work has been done, and we know very little indeed about them; as we shall see, however, in the general context of topographical mapping in medieval Europe they may at least equal the other two groups in their interest and significance.

The oldest of the district maps to survive from medieval Italy occurs in a manuscript dated 1291; it is very badly damaged, but enough survives to show that we have a nearly identical later copy of it preserved entire in a manuscript of the second half of the fourteenth century (it cannot be older than 1353). The map shows an area south-east of Turin, around Alba and Asti, and it has south at the top. The two

towns are shown as rectangles, but other settlements – over 160 of them – are marked with little castles, towers and flags and coloured red. Scale is inconsistent, and the relative positions of the places shown are only roughly correct. Another fourteenth-century map – it can be dated to the last twenty years of the century – is of Lake Garda. This shows the lake-shore settlements and, significantly, the fortifications there with great exactness, clearly from actual observation.

No other of the surviving district maps is older than the fifteenth century, though there are contemporary references to a map of Padua and the surrounding area drawn by Jacopo Dondi, who died in 1359, and also to a map of all Lombardy that was made for the lord of Padua, Francesco da Carrara, in 1379. But from the first half of the fifteenth century we have a rather rough sketch map of the area around Brescia and two maps of Lombardy. One is dated 1440 and is signed by 'Ioanes Pesato', who is otherwise unknown; the other is unsigned and undated, but it has been suggested that it dates from the war that was fought between Milan and Venice in 1437–41. Although there are similarities in general style and in some details, the two are quite different maps. Thus Pesato's map has west at the top; the anonymous map does not extend as far east as Pesato's and is oriented to the north, having as its horizon the line of the Alps, broken only by the towers of some of the towns at their foot, such as Como and Lecco. The anonymous map marks more places than Pesato's, and its representations of walled towns are elaborate and varied, but Pesato's map shows the larger towns more carefully and more accurately. On both maps distances along roads are marked, and in some cases, but not all, the figures given are the same on both. The date on Pesato's map clearly places it in the period of the war; the reason why the anonymous map is assigned to the same period is its interest in features of military significance. Thus walled towns are distinguished from undefended ones; river crossings are clearly marked, and separate, readily identifiable, pictorial signs are used for wooden bridges and for stone ones; in the mountains is shown the 'Fossa Bergamasca', the boundary ditch that was dug between the lands of Milan and Bergamo in the thirteenth century. The map forms, in short, a graphic guide to the

29

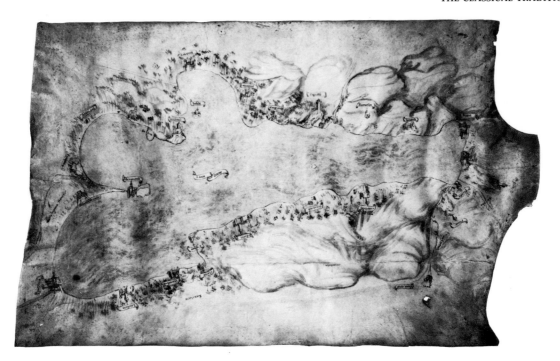

29 Map of Lake Garda, late 14th century. One of the earliest of the late-medieval group of district maps from north Italy. Like many picture-maps it has no right way up: most of the wording and pictures simply face inwards to the lake.

theatre of war at the time of the first Venetian conquests on the *terraferma*, inland Italy.

Another map which has been dated to the same war period but which may be rather later – opinions have differed – is of Verona and the surrounding area. Verona itself appears in the centre, drawn on a far larger scale than the rest of the map and with a wealth of detail: its churches, bridges and fortifications all appear pictorially with recognizable accuracy. The roads radiating from the town are a prominent feature of the map, and beside the principal ones are tables giving the distances of places from Verona. Other towns and villages are also shown in perspective, in greater detail the nearer they are to Verona, and walls and fortifications are clearly marked even in small places. Colours are naturalistic: roads are yellow, vegetation light green, rivers and lakes blue-green, and mountains are shown in a brown that is intensified to

indicate greater height. The forms of place-names suggest that the map was made locally, and the artist must have known the area well himself – he was clearly familiar with the appearance of the mountain groups along the Adige valley. Other details on the map include logs drawn beside a river to show that it was used for floating them downstream, and a picture of six ships being transported overland from the River Adige to Lake Garda – a reference to an incident of 1439, when the Venetians got ships to the lake this way so as to relieve Brescia from siege by the Milanese. These details are peculiar to this particular map, but some of its more basic features recur on other district maps of the same period; these too have a town, exaggerated in size, at the centre, roads prominently marked, giving distances, and a special emphasis on fortifications. The last two features, of course, are shared also with the two maps of Lombardy. These other district maps are of the areas around Padua (1449), Parma (not earlier than 1460) and Brescia (1471–2 or possibly earlier). The first is by Annibale di Maggi, the others are anonymous; older models may have been used in drafting the maps. The Brescia map shows the city's fortifications in

IV

59

detail, but the area inside the walls is blank, while other settlements are shown by a wall with towers or by a house or a church. The Parma map shows the chief buildings in the city, and marks other settlements by pictures that only partly correspond to their actual appearance.

Of the four cities at the centres of these maps all but Parma belonged to Venice when the maps were made. In fact when we look at the district maps from north Italy we find ourselves, in the course of the fifteenth century, turning more and more towards Venice: they relate to Venetian wars, to Venetian territories of the *terraferma*, and we may reasonably wonder whether it was from Venice that they at least drew their inspiration. Fifteenth-century Venetians were certainly aware of maps. As we shall see in the next chapter there was a tradition of plans of the city, and in the late 1470s maps of the world and of Italy were painted on the walls of a room in the Doge's palace – but all this can be paralleled in other cities of medieval Italy. What seems to have been peculiar to Venice was its appreciation of the role maps could play in administration and, particularly, in military planning. In 1460 the Council of Ten ordered the governors of all territories, cities and castles under Venetian rule to have maps made of the areas within their jurisdiction and to send them to Venice. Three surviving maps are thought to have originated in this decree, though they are all of later date and it must be assumed that revised or improved maps were demanded from time to time in succeeding years. The oldest is of the territory of Padua, produced in 1465 by the Paduan painter Francesco Squarcione (Squazòn in the local dialect). It is very similar to Maggi's map of 1449, though it is rectangular instead of circular in shape. Another is of the territory of Brescia, covering Lakes Iseo and Garda and some 280 named places. It has been dated to 1469–70, and would thus be slightly older than the map of Brescia and its surroundings that has already been mentioned; but it is similar at least in having the city at the centre of the map and of a radiating road system. The third is another map of the territory of Verona, dating from 1479–83; there is some overlap with the Brescia map (they both show Lake Garda, for instance) and if they really do originate in government orders for maps of the territories, their bounds

were not strictly observed. Both these last two maps are extremely complex – bridges and mills are among the many details shown. Finally we have a map of the whole of the Venetian *terraferma*, together with some other areas, for it extends west as far as Como, south to Mantua. Its date is uncertain, but it is probably not later than 1496–9, as the bird's-eye view with which it marks Venice itself lacks the clock-tower built during those years. Other towns are marked by fortified walls which in some cases, such as Verona, portray the actual layout of the defences. It can be seen as primarily a military map, showing the fortifications of the Venetian territories, and the forms of its place-names show that it originated at Venice. In view of this its present location is particularly interesting, for it is in the Seraglio Library at Istanbul, and early transliterations of its principal place-names into Cyrillic script suggest that it had left Italy at an early date. Clearly not only the Venetians but also their Turkish enemies appreciated the value of maps; the Turks are known to have used portolan charts of Italian origin.

But we must not overemphasize the importance of Venice in the tradition of north Italian district maps. After all, the earliest known, the map of the area around Asti and Alba, comes from the opposite end of the River Po, and it has been suggested that several of the fifteenth-century maps may have been based on antecedents dating from before the spread of Venetian power. And whereas the earliest detailed map of Switzerland, drawn by Konrad Türst in the 1490s, almost certainly owes something, possibly a great deal, to the north Italian district maps, Türst's Italian contacts were not with Venice but rather with Milan. On the other hand it cannot be just a coincidence that most of the surviving maps have Venetian associations; it is not due to chance preservation of a single cartographic archive, for the individual maps have varied histories and are now in different repositories throughout the region and beyond. What seems most likely is that the rulers of Venice in the fifteenth century, realizing the value of maps in planning military campaigns and in defending and governing their newly won provinces, developed and put to full use a tradition of map-making that had existed in the north Italian plain for at least a

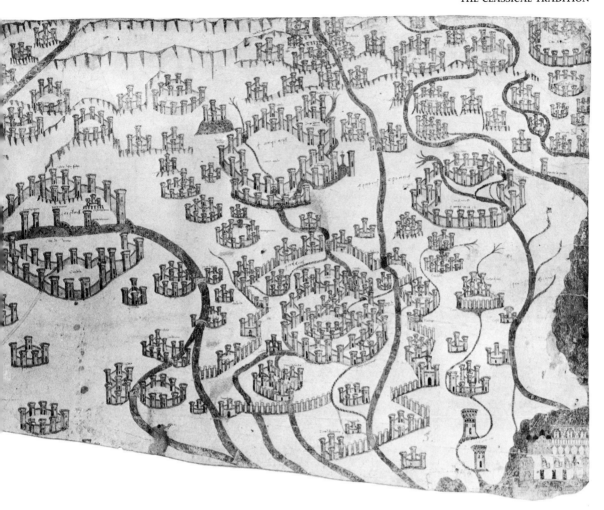

century or so. In this they were well in advance of their time. We know of no other state in fifteenth-century Europe that used maps in the work of government. It is of very great interest that we have here not a general growth of the use of maps among several neighbouring states, or even more widely, but a single state that recognized their potential and developed them in isolation on the basis of an existing local tradition.

But, setting aside the possible role of Venice, it is of no less interest that all these district maps from medieval Italy come from this one region, the north Italian plain. Far more map-making was done in Italy than anywhere else in medieval Europe, but even here it was a matter not of a widespread knowledge and use of maps, but rather of traditions of particular sorts of map

30 Highly stylized map of the Venetian *terraferma*, late 15th century. Venice is shown, bottom right, by a picture of St Mark's Square and the Doge's palace; other towns include Verona (centre left) and Padua (lower centre). At the top irregular lines with deep indentations mark the Alps.

(whether geographic or topographical) in particular contexts. The tradition of district maps was peculiar to north Italy – a point we shall return to when we look at similar pockets of local map-making in other parts of medieval Europe. Certainly elsewhere in Italy there is nothing comparable. A map of the Straits of Messina appears on a seal of the Emperor Frederick II that was used in 1226; it provides

61

an odd echo of the Zancle coins, for it too shows plainly the distinctive sickle-shaped sandbank. In 1390 a painter named Ambrogio Benincasa was paid for making a map, which has not survived, of the country around Florence and Siena. A map of Tuscany by Pietro del Massaio appears in a manuscript of Ptolemy's *Geographia* that is dated 1456, and similar maps of 1469 and 1472 derive from it; it is a map of great interest and significance, not least for its advanced style (it has no pictorial elements), but it belongs in part to the tradition of geographic maps, an expansion inland of the coastal maps provided by the portolan charts. Equally significant in a different way is a map of the northern boundary of the kingdom of Naples, which we know only from a late copy. Its title names its author as Giovanni Pontano and says that it was made on the orders of King Ferdinand; there was a Ferdinand on the Neapolitan throne most of the time between 1458 and 1516, but the map probably belongs to the end of this period for – and this is what makes it especially interesting – it is based on a measured survey and is specifically drawn to scale. We shall be referring to it again in chapter 9 with other surveyed maps. There is also some reason to suppose that there was an early general map of the kingdom of Naples in the archives at Versailles in the eighteenth century (if so it is now lost). But all in all there is no evidence for a counterpart elsewhere in Italy to the north Italian tradition of district maps.

We know so much about the Italian district maps of the late middle ages because of the work done on them over the last hundred years by a number of scholars, above all by the late Roberto Almagià. But when we pass to the next group of topographical maps from medieval Italy – the maps and plans of particular localities and plots of land – the situation is completely different. So few have been reproduced or even recorded in print that it is no more than a guess that a substantial number of maps of this sort exist at all. But the guess is likely to be a good one. For one thing it would be thoroughly in keeping with what has happened in other parts of Europe for this type of material to be overlooked. As already suggested, historians of cartography have been a little apt to assume a readiness to draw maps. Accordingly no one would think it worth while to draw particular attention to this or that instance where someone made a sketch-map to illustrate the position of his field or vineyard, to show the layout of houses in a single street or village, or to back up his contention in a case over landed property or other rights in a court of law. These were things that anyone might do in a culture as advanced as medieval Europe's. This is a far from safe assumption, and we shall be looking at it critically when we examine the relevant evidence from other parts of Europe in chapter 5. But we have a particular reason for thinking that medieval Italy produced local maps of this sort, because in 1355 a treatise by an eminent Italian lawyer and legal theorist, Bartolo da Sassoferrato, used diagrams as a way of resolving disputes over watercourses. We shall see how some local maps from fifteenth-century France can be clearly connected with this treatise, and it seems likely that if it exerted influence there it would equally have affected practice in the author's native Italy. But if there are maps of this sort lurking in the medieval archives of Italian law-courts and estate-owners they have yet to be brought to light and discussed in print.

However, there are at least four instances where Italian local maps of this sort have been recorded. One map is of the Chiano valley in Tuscany and dates from the thirteenth century. It was discussed in the early nineteenth century by Count Fossombroni for the light it throws on early connections between the river-systems of the Chiano and the Arno, and his study, with an engraving of the map, was published in 1824 in a collection of works by Italian authors on hydrography. It is from this reproduction that ill. 31 is taken. Another, dated 1306, is of a site with a harbour at Talamone on the Tuscan coast which the city of Siena had bought three years earlier with a view to building it up as a town; the plan shows plots laid out for settlement and gives the names of the people who were going to occupy them. Others are a sketch, part map, of the lagoon area around Venice and a small group of maps, drawn in ink, of areas around Ravenna; all these are of the fifteenth century. It may be that we know of few maps of this sort from medieval Italy simply because few exist; it seems more likely that those we know are the merest tip of an iceberg. They deserve far fuller investigation than they have yet had.

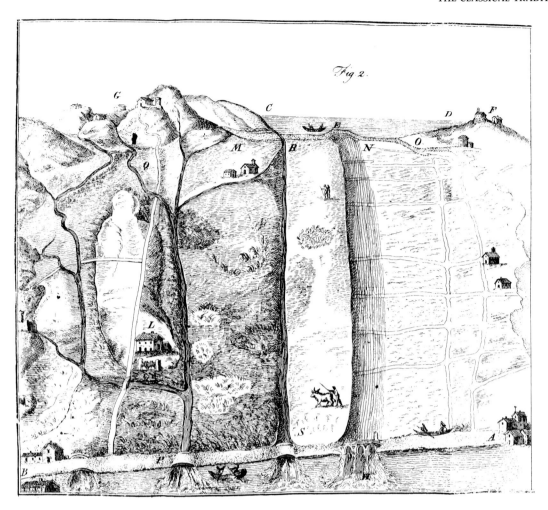

31 Very few local plans are recorded from medieval Italy, though many may remain to be discovered. This, showing part of the Chiano valley near Arezzo and originally in the archives of a monastery there, is reproduced from an engraving of 1824. The letters on the map are the engraver's additions, but he succeeds in conveying the general character of a medieval picture-map.

Before we pass in the next chapter to the third group of Italian topographical maps – the plans and bird's-eye views of cities – we should digress for a moment to look at a different aspect of the cartographical output of medieval Italy. We have already seen that there were Italian maps of the Holy Land; there were also maps of other places outside Italy. Some, like the portolan charts, lie outside any tradition of topographical mapping. But others are like Del Massaio's map of Tuscany in 1456; they lie on the border between the topographical and the geographic map. We are, after all, approaching the point when the distinction between the two began to be blurred. But there is no difficulty in discerning the elements of the picture-map in their ancestry. Thus a map of 1453 showing the lower Danube with its delta has pictures of castles (to mark towns), mountains, trees and a bridge over a tributary stream. Particularly interesting are the books of islands. These seem to have been first compiled as an aid to navigation, supplementing the portolan charts: they were collections of plans of islands and

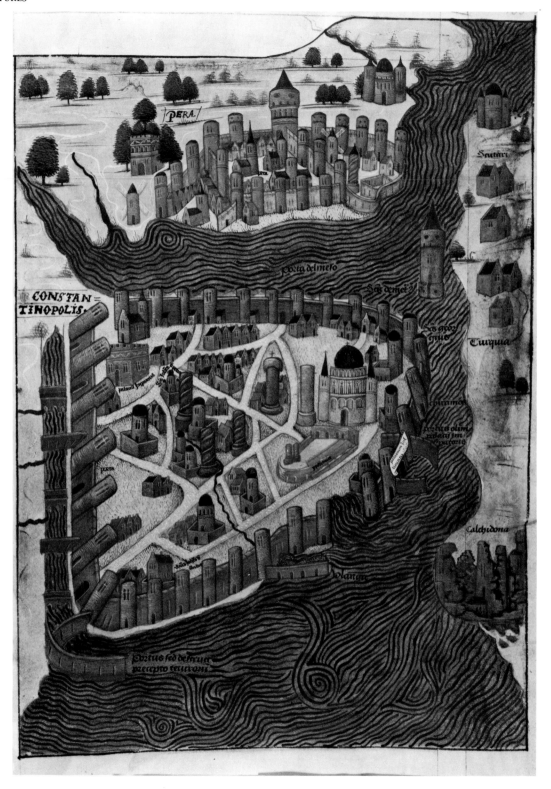

harbours in and around the Mediterranean. But the more sumptuous versions that survive were obviously not for use on board ship. The oldest we have is by Cristoforo Buondelmonte and dates from 1420; one by Bartolomeo dalli Sonetti was printed at Venice as early as 1485. Their plans include bird's-eye views of towns which are based on their actual appearance; some even name principal buildings, as we see in Buondelmonte's map showing Con-

32 stantinople. This map varies a good deal from one manuscript to another of Buondelmonte's work; this reflects the book's popularity – it was copied often and widely – and the interesting fact that artists copying the map were apt to incorporate in it information drawn from other sources. A precursor of the book of islands is a map of Egypt, with little perspective views of towns, that was painted about 1360 by Gerardo Starnina and is now in the Uffizi Gallery at Florence.

But the methods of the picture-map need not be used only for depicting real places. As we have seen, imaginary or conventionalized pictures appear on these maps as well as real views. The pictures of towns, castles, forests and mountains might bear little relation to their actual appearance or none at all; they merely marked the presence of a particular type of settlement or landscape. Bird's-eye views – picture-maps in effect – might be made of quite imaginary landscapes simply as works of art. In the Pinacoteca Nazionale at Siena is a pair of small paintings by Ambrogio Lorenzetti, who worked in the first half of the fourteenth century: both are landscapes seen from above, and both could be called picture-maps. But neither scene has been identified, and probably both are imaginary; one shows a city by the sea, the other a castle, also beside the sea, with a lake behind. In the same gallery is a picture of Bernadino of Siena, the Franciscan preacher and saint, painted in 1450, only six years after his death, by Sano di Pietro. At the saint's feet is a symbolic representation of the earth; its form is that of a picture-map, a multitude of small islands with bird's-eye views of buildings and trees. The picture-map's technique of showing landscape was in fact an accepted artistic convention of late-medieval Italy. This is something that we should bear in mind when we turn to the plans and bird's-eye views of towns.

32 Constantinople as shown in a 15th-century manuscript of Buondelmonte's book of islands. North is at the top. The city itself is in the foreground; beyond is Pera on the other side of the Golden Horn. The channel to the Black Sea is to the top right. A few buildings are named, among them the emperor's palace ('palacium Imperatoris') in the north-west corner of Constantinople.

4

Town plans and bird's-eye views

WE HAVE just been considering some Italian medieval paintings which show landscape viewed as from a height above the ground. It may have seemed arbitrary to distinguish these from others showing views from ground-level, and when we turn to representations of towns in the middle ages the distinction may seem even more artificial. In the illustration we see part of one version of the map showing the route from London to Apulia that was drawn by Matthew Paris, the thirteenth-century St Albans monk. If we look at the little pictures of walls and buildings marking the towns we see that some, like Dover, are drawn in elevation as seen from the ground, others, like Canterbury, obliquely as though from a height. All are sketchy in the extreme; in a few cases one or two identifiable structures are drawn in or even named, as St Paul's Cathedral in London or St Augustine's Abbey at Canterbury, but most of them are drawn as conventional or entirely imaginary views. It might seem quite unreasonable to describe some of them as picture-maps (in this case, of course, maps within a map), the others as ordinary views, as they all belong so palpably to a single style of draughtsmanship and to the vision of a single artist. When we move to an altogether different type of town view the distinction may seem no less absurd. The finest of the medieval representations of Florence, the so-called 'Map with the chain' dating from the 1480s, takes the form of a simple landscape view looking down from hills to the south-west of the city; in the foreground, as if underlining this by a flight of whimsical fancy, we see the artist, sitting on the ground with his sketch-book, actually drawing the picture. In fact this is deliberately misleading. There are no hills overlooking Florence at the point from which the picture was supposed to have been drawn; it

shows the city from a viewpoint unattainable in reality and is a cleverly constructed picture-map. But is this not the merest quibble? Florence, after all, is nearly surrounded by hills; if there had happened to be one at the supposed vantage-point of the 'Map with the chain' and if the map had actually been drawn there from nature the result would probably have been more or less exactly what we now have, yet we would not be calling it a map at all. Why should an oblique view of a city or landscape in the plain be regarded as a picture-map while a similar view of an Alpine valley, as actually seen from the overlooking mountains, lies outside the history of topographical mapping?

But these are, of course, difficult cases; and whereas difficult cases do not necessarily make bad law, they put good law to its severest test. In fact our definition of the picture-map stands up to them well, even if it seems to operate arbitrarily. Matthew Paris's representations of Canterbury and Dover may look very alike apart from the angle of vision, but in this apparently slight variation there lies a great difference in concept and probably also in tradition. And by the time of the Florence 'Map with the chain' bird's-eye views of cities had evolved methods and techniques that distinguished them clearly from straightforward ground-level pictures. If there had been a hill just where it appears in the foreground of the 'Map with the chain' and if the city had been drawn from it, the result would indeed have looked very like the view we

33 First part of Matthew Paris's itinerary from London to Apulia, mid-13th century. On the left is the route from London to Rochester, Canterbury and Dover, on the right alternative routes from Ouissant ('Witsaint') through Poix and Beauvais and from Calais through Arras and Rheims.

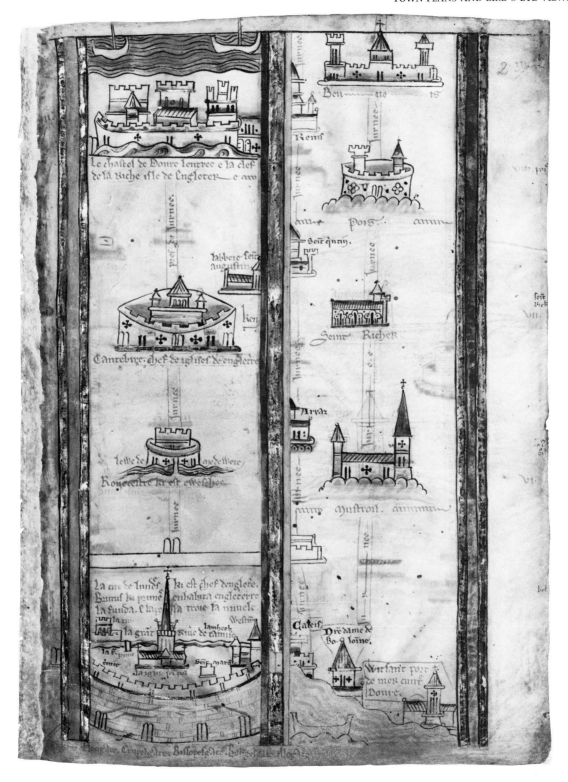

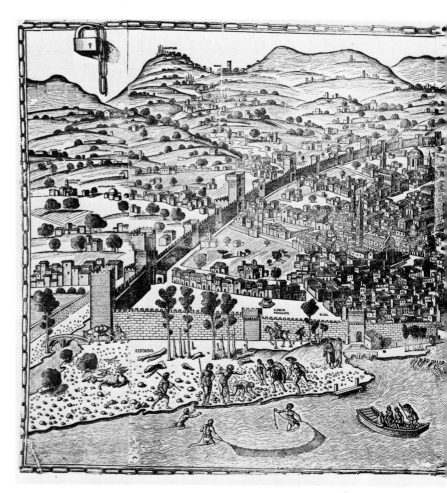

34 Apparently a simple view of Florence from a hill above the city, in fact a cleverly constructed picture-map: the hill in the right foreground does not really exist. Known from its border as the 'Map with the chain' (note the padlock top left), it is a reduced woodcut version of a much larger bird's-eye view of Florence, published by Rosselli about 1482, of which no copy survives.

actually have, but it would have been arrived at by quite different methods of draughtsmanship. It is still not certain exactly how late-medieval artists set about drawing detailed and reasonably accurate pictures of such complicated subjects as if from impossible viewpoints; but the development of the necessary techniques underlies some of the more important changes to be described in this chapter. We shall be concerned here almost entirely with plans and views of individual towns drawn by themselves; this leaves untouched an important part of this topic, for we have many more plans and views of towns on medieval maps of larger areas, from the little sketches on Matthew Paris's route map to the detailed plans of towns on which the north Italian district maps are centred. But to do them anything like justice

would demand not just another chapter but an entire treatise. They have never been investigated systematically; those on some maps have been fully analysed, but on many others there is still much patient topographical and historical research to be done. When a general picture can be built up of their pattern of development it may well prove significant and revealing. But meanwhile the plans and bird's-eye views produced of single towns by themselves will serve to show the main lines of development and the particular problems of this type of picture-map; those drawn as features on other maps can be left for the more detailed investigation that they deserve.

One question that arises at once, as it did with the other picture-maps from medieval Italy, is whether we are dealing with a style of

representation that originated in the middle ages or whether there was a continuous tradition of plans and views of towns from the Roman past. In fact for city views it is easier to find what are likely to be continuing threads of tradition than it was for the other types of picture-map. One of these can be seen in the way towns are shown in the Roman surveyors' treatises; it was probably very widespread in the Roman period. If we look at the treatises' map of the area around Terracina we see that the town itself is shown simply by a picture of its walls, viewed from above so that the ground inside can be seen. When other identifiable towns are shown on the maps in these treatises the shape and other details of the walls vary, perhaps reflecting the actual features of the towns' fortifications, and sometimes one or two

buildings are shown inside the walls. But elsewhere in the surveyors' treatises, where we have not a map of a particular place but a stylized picture or diagram to illustrate some point of theory or technique, we find towns shown in similar form but in a purely conventional way, the walls forming a regular hexagon or other polygon with a tower at each angle. It is this stylized view of a town – what Professor Lavedan has called a city ideogram ('l'idéogramme urbain') – that we can trace in a variety of contexts from Hellenistic art through the Roman period and down to the high middle ages. It is used to represent Athens and Corinth on a cup of the second or third century BC from Tanagra in Boeotia. It occurs in the Vienna Gospels of the fifth or sixth century, in sixth-century mosaics at Gerasa (now Jerash in Jordan), in the ninth-century Utrecht Psalter. It was then normally hexagonal, but later it became octagonal or circular, as in the stained-glass windows at Sens and Chartres. Sometimes buildings are shown inside the town walls, sometimes figures or scenes to fit the particular illustration, but the walls with their towers, viewed obliquely from a height, are constant. It is to this tradition that the oblique views on Matthew Paris's route map belong; he was one of the first medieval artists using this form to distinguish between one town and another by drawing and naming their prominent buildings, but he was anticipated by the seal of the Emperor Frederick Barbarossa (1152–90), which represented Rome by combining the city ideogram with a picture of the Colosseum. These city ideograms are the clearest link we have between the picture-maps of Roman antiquity and those of the middle ages.

The possibility of a more specific, if more tenuous, link is suggested by Einhard's life of Charlemagne, the first biography of the emperor, written in the early ninth century by someone who had known him intimately. This tells us that Charlemagne at the time of his death had among his treasures panels ('mensae') of silver and gold engraved with representations of Rome, Constantinople and the whole world; Einhard does not tell us whether they had been newly executed or were already antique. It has been suggested that the one of Rome may have been based on a surveyed map of the city, like the only map of Rome that we have from classical antiquity, the vastly detailed third-century plan, carved on marble, that we shall discuss in chapter 8. Its being engraved (though on silver, not marble) might be held to lend colour to this suggestion; and it is perfectly possible that there once existed similar plans from which the panel showing Constantinople might have derived. In this case they provide no link with later maps: there is no conceivable connection between the surveyed map of ancient Rome and the medieval picture-maps of the city. But it seems just as likely that Charlemagne's panels bore picture-maps, elaborate versions of the sort of oblique view which we find in the surveyors' treatises and which seems to have been the normal form of map used by the Romans for non-technical, illustrative purposes. It may be that the earliest medieval bird's-eye views of Rome are descended from

picture-maps of the city drawn in classical antiquity.

However, they are far more likely to have taken their immediate inspiration from the plans of Jerusalem and other towns in the Holy Land that date from the period of the Crusades. Given our ignorance of early topographical mapping in Islamic countries it may be that the crusaders got both the idea and the contents of these plans from existing local maps in an Arab tradition. But here too a direct line of descent from classical antiquity is possible. The mid-sixth-century Madaba mosaic includes very respectable III picture-maps of Jerusalem and other towns. Another plan of Jerusalem accompanies some copies of the account of the pilgrimage that the Gallic bishop Arculf made about the year 670. In form it is little more than one of our city ideograms, but it provides some topographical information in naming the city gates and in noting the way to the valley of Jehoshaphat. The centuries that followed produced other travel books by pilgrims to the holy places of Palestine, but we have no more plans of its cities until the crusaders' capture of Jerusalem in 1099. Then, from the first half of the twelfth century on, we have a series of plans of Jerusalem.

One of the most accurate in its detail is also 35 one of the earliest; it must be based on actual observation and it dates probably from the 1140s. Although very much more than a city ideogram it echoes the same technique: the city walls, gates and towers are shown in full, but inside the walls we see only the main streets, a couple of hills and the holy places and churches. The point where the crusaders broke into the city is marked ('Here the city was taken by the Franks') and on the buildings we see the belfries that were introduced to Jerusalem in the early twelfth century; sometimes a belfry is shown where none was built, but this may well have been anticipation of expected development, not ignorance or carelessness on the part of the

35 Dating from the 1140s, this plan of Jerusalem shows real knowledge of the city, which had been in the Crusaders' hands since 1099. Top left we see the court of the king ('curia regis'), that is of the Crusader king of Jerusalem.

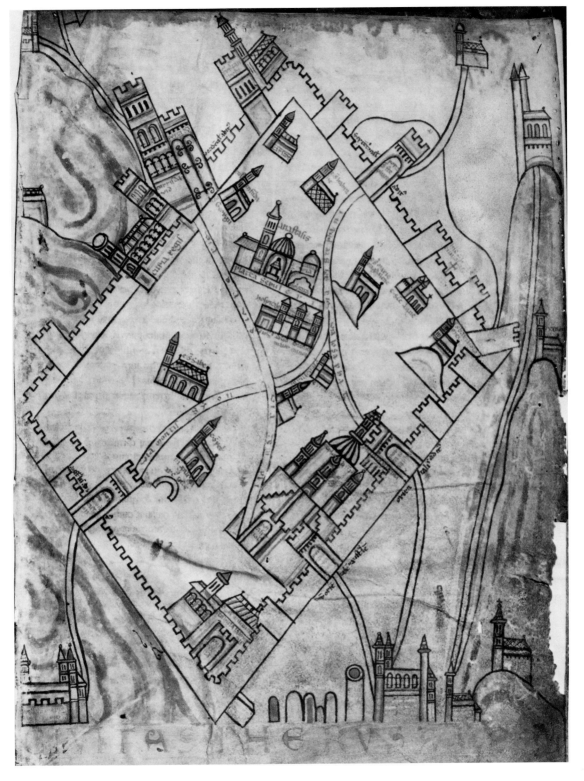

artist. Most of the medieval plans of Jerusalem, though likewise showing real knowledge of the city's layout, are more diagrammatic: the walls are circular and the street-plan rectangular. They are clearly related to one another, but at the same time there are many differences of detail between them and their exact relationships have never been worked out. It is in this group that there are most likely to be links with a late-Roman tradition of plans of Jerusalem. It is in a version of this style of plan that we find Jerusalem appearing in the centre of the elaborate medieval world-map at Hereford Cathedral. Of later medieval plans of Jerusalem the most outstandingly accurate are those that accompany the early-fourteenth-century works of Marino Sanudo and Paolino Veneto: Sanudo's *Liber secretorum fidelium crucis* and Paolino's *Magna chronologia*. Both authors were native Venetians and both had distinguished careers, Sanudo as a diplomat and Paolino in the church (he became Bishop of Pozzuoli). Sanudo wrote as a propagandist, urging a new crusade, Paolino as a historian – his work is a world history. Both books are illustrated with geographic and topographical maps. Those in Sanudo's work may all (like his map of the Holy Land mentioned in chapter 3) have originated in the workshop of Pietro Vesconte at Venice; Paolino's book reproduces some of these, but adds some others as well. Thus most surviving copies of both books include maps of Acre as well as of the Holy Land and Jerusalem, while all copies of Paolino's work also contain one of Antioch. We have already seen how one of Matthew Paris's two maps of the Holy Land is dominated by its plan of Acre. Thus we find plans of towns in Palestine being drawn both during the crusading period and after, and in various parts of Europe: the Jerusalem plan of the 1140s was probably drawn in Lorraine or Flanders, Matthew Paris worked in England, manuscripts of Sanudo's and Paolino's works were copied at Avignon. But it was only in Italy that we find any local tradition that may have derived from them of drawing plans of towns.

The medieval plans and bird's-eye views of Italian cities should be seen then against this threefold background: the age-old tradition of the city ideogram, a possible ancient tradition of picture-maps of Italian towns, and the plans of towns in the Holy Land that began to appear after the first crusade. Any or all of these may be connected with the picture-maps of towns in Italy that were drawn from the twelfth century on. Enough of these survive for us to be able to speak confidently of a genuine tradition, but all the same they are not very numerous. Rome appears far more often than any other city, but of Rome we have only some twenty representations, based on six or seven prototypes. The oldest probably dates from the twelfth century but it is known only from a fourteenth-century copy. It is more of a symbolic diagram than a map: towns and battlements mark the city walls, a line below them the River Tiber. Thirteen individual sites are named, some with pictorial symbols, but their arrangement owes little or nothing to their positions on the ground. It seems to show particular interest in places connected with St Peter, and it may have been first drawn to accompany a descriptive guide to the city, *Mirabilia urbis Romae* ('The wonders of the city of Rome'). Our next plan of Rome is much more elaborate. From the buildings shown it can be dated soon after 1280 and it may have been drawn in the first place to illustrate an enlarged version of the same guide. We know it only from its inclusion in three of the four known manuscripts of Paolino Veneto's *Magna chronologia* which date from the 1320s and 1330s; unlike the same book's maps of the Holy Land, Jerusalem and Acre it is not to be found in Marino Sanudo's *Liber secretorum*. The basis of the plan is a full outline of the walls (all the principal gates are named) and inside we have the line of the Tiber, the hills and a profusion of the city's monuments shown in elevation – we can easily distinguish the aqueduct, the Colosseum, Castel Sant' Angelo and many others. But the artist has also taken a first step towards naturalistic representation, towards showing what the city really looked like, by filling the blank spaces between the

36

36 Plan of Rome known only from copies of the 1320s and 1330s but probably composed some forty years before. Prominent are the city walls (with the gates named), the course of the Tiber and the seven hills with buildings on each. It is a spirited representation but far from realistic. The foot of the map is approximately north-west.

35

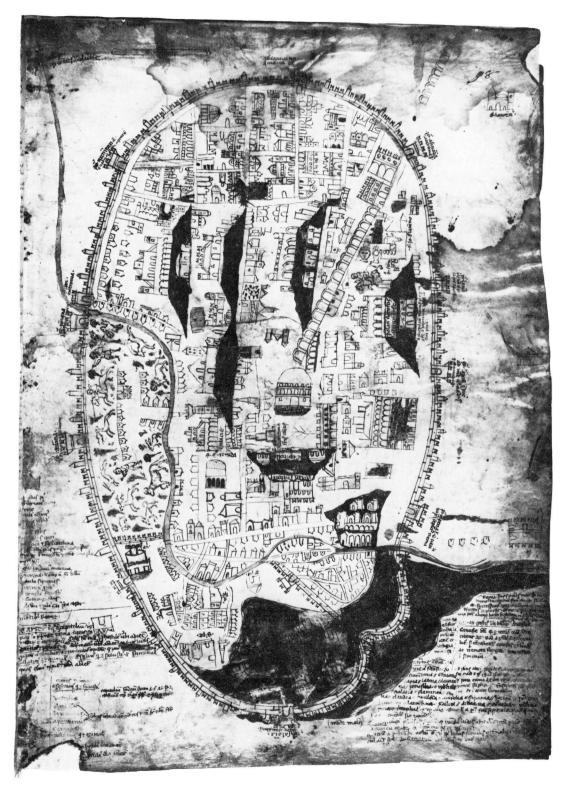

37 Seal of the Emperor Ludwig the Bavarian, designed for seals struck in gold. It was first used in 1328, the year of his coronation at Rome, and the monuments representing the city may have been chosen to refer to this and to other events of his stay there. No fewer than twelve identifiable monuments appear, dominated, top centre, by the Colosseum.

monuments with rather impressionistically drawn streets lined with houses. And on the north-east side of the city we see a stag-hunt in progress.

But until the late fifteenth century there was no further advance in the form of these maps: the outline of the city's walls and pictures of its most striking monuments form the pattern for subsequent plans of Rome. These occur in very varied contexts. One is on the seals that were struck in gold for the Emperor Ludwig the Bavarian from 1328 on; in small compass this gives us, as Professor Juergen Schulz has put it, 'a veritable compendium of the monuments of the Eternal City'. Another occurs in the mural paintings executed in 1413–14 by Taddeo di Bartolo in the Palazzo Comunale at Siena, while a version of the same plan (of which the lost prototype dated from before 1348) was included in the highly illuminated book of hours known as *Les très riches heures* produced a few years later for the Duc de Berri by the brothers Limburg, working in France. This form of map particularly suited the interests of the more

archaeologically minded of the fifteenth-century humanists; one is connected with Flavio Biondo's account of Rome's antiquities, *Roma instaurata*, which was written in 1444–6 and remained for a century the principal guide to its ancient monuments. This map showed simply thumb-nail sketches of churches and of classical buildings and ruins inside and outside the walls, giving them the names assigned to them by Biondo. Again it is known only in copies of later date than the original: one drawn in Venice by Alessandro Strozzi in 1474 and three others included in manuscripts of Ptolemy's *Geographia* from the workshop of the Florentine Pietro del Massaio in 1469, 1472 and the 1490s. The map's features reflect the interests and aspirations of the Renaissance, but in its cartographic form it belongs unquestionably to the middle ages.

We find a similar pattern in the picture-maps of other Italian cities in the middle ages, though of none do we have so many surviving as of Rome. An early example is a plan of Milan, shown in circular form with little detail, that accompanied the chronicle written by Galvano Fiamma about 1330. The earliest plan of Florence is on a fresco in the Loggia del Bigallo there and dates from its construction in 1352–8. It shows, simply, some of the principal monuments of Florence: the cathedral, the baptistery, the Palazzo Vecchio and a very few others, among them Giotto's Campanile beside the cathedral, shown in course of construction, as it was at that time. Another plan of Florence, like that of Rome, was included by Pietro del Massaio in manuscripts of Ptolemy's *Geographia*; it shows the city walls, the River Arno and principal buildings with their names, all very competently drawn, but it gives no indication of other buildings or of the pattern of streets. Plans of Milan and Volterra are among other maps included in Massaio's Ptolemy manuscripts. Far more of these stylized plans or bird's-eye views of Italian towns occur on fourteenth- and fifteenth-century maps of larger areas, showing just how well this particular form of representation was established in medieval Italy. They range from some of the elaborate plans on the district maps from north Italy described in the last chapter to the tiny sketches that sometimes represent Venice and Genoa on portolan charts. We find too the same technique used by Italian artists to show places outside Italy. We have already seen that picture-maps of Constantinople and other towns occur in the fifteenth-century books of islands, and Massaio's Ptolemy manuscripts include plans not only of the Italian cities, Jerusalem and Damascus but also of Constantinople, Adrianople, Alexandria and Cairo.

But by the late fifteenth century this form of plan, showing towns by little more than pictures of their walls and their most notable buildings, had become quite out of keeping with the trend towards realism in contemporary Italian art. Much earlier than this we find ground-level views of Italian cities showing them not in near-symbolic form by a few key features, but in full detail as they really were. It is thus that we see Siena in the fresco symbolizing good government that was painted in the Palazzo Comunale there by Ambrogio Lorenzetti in 1337–9 (an odd contrast with the plan of Rome painted by Taddeo di Bartolo in the same building three-quarters of a century later); it is thus again that we see Perugia in the paintings by Benedetto Bonfigli commissioned by the city in 1454 for the chapel of the Palazzo dei Priori. Clearly in a ground-level representation, where the artist had simply to reproduce what he had before his eyes, it was much easier to introduce this detailed realism than in a bird's-eye view which would involve the elaborate construction of just what would be seen from an unattainable vantage-point. A detailed view of Padua, as from a slight elevation, in a fresco painted by Giusto de' Menabuoi in 1382, represents an early attempt to meet this problem. Certainly by the late fifteenth century appropriate techniques had been developed and there started to appear a new form of picture-map of Italian cities: bird's-eye views that set out to show what a bird really would see.

A name particularly associated with this development is that of Francesco Rosselli, an engraver of Florence. Oddly, there now survives no first-hand specimen of his work in this genre, and his connection with it rests on four entries in an inventory, drawn up on his son's death in 1525, of the contents of the workshop that had once been his. This inventory lists the printing blocks and plates for large-size views of Pisa (woodcut), Florence, Rome and Constantinople (all three engraved). Of the views of

Pisa and Constantinople we know nothing, but the other two survive at second hand in views that are believed to derive from Rosselli's engravings. Of the view of Rome the fullest version is a large painting in the Ducal Palace at Mantua; it cannot be earlier than 1534, but Rosselli's original probably dated from the 1480s. It shows Rome as though from a considerable height, necessarily so in view of the sheer size of the area encompassed by the city's walls. Ancient monuments are still very prominent – the Colosseum and the Pantheon dominate the left and right sides of the view – but they appear in the context of the whole city as it was at the end of the fifteenth century. Of Rosselli's view of Florence we did have, until its destruction in the Second World War, a single impression from one of the six original plates which were probably engraved about 1482. Derivative views, of the 1480s and 1490s, survive in a painting and, reduced in size but fully detailed, as a woodcut, the 'Map with the chain' that we have already seen. Rosselli's Florence is drawn as from a lower height than the view of Rome, and it is worth noting that it differs from a straightforward view in that some of the buildings, among them the Palazzo Vecchio, have been so to speak turned round to present their most distinctive façade to the observer – a technique inherited from the older style of town view. But Rosselli's views were not the only ones to be produced in the new style of detailed realism. Beside them can be placed paintings giving two versions of a view of Genoa, and perhaps also a painting of Naples, though here the angle of elevation is slight; in each case the original view must have been drawn in the 1480s. There is a printed view of Ferrara dated 1490. And, above all, there is the view of Venice by Jacopo de' Barbari.

This extraordinary work was published in 1500: the date is given in its title, 'VENETIE MD', and is confirmed by other sources. It was printed from six woodcut blocks and is of great size, measuring some 4 by 9 feet (1·3 by 2·8 metres); it was very much larger than Rosselli's view of Florence, which was about 2 by 4 feet (0·6 by 1·3 metres) and probably somewhat larger than his view of Rome. What we see here is no more than a part of the centre of the view, but it serves to give an impression of the great detail and complexity of the whole. The city

34

38, 39

32

appears as from the south-east and at a considerable height for, as in Rosselli's view of Rome, the artist had a large area to cover. At the foot are the islands of Giudecca and San Giorgio Maggiore; the whole of Venice and Murano are shown in full detail and in the background are Torcello and other islands of the lagoon, the mainland shore and, forming a distant horizon, the long line of the Alps. We see the city in the low eastern light of early morning. In the harbour there are many sea-going ships, and everywhere there are smaller boats to be seen. Also in the harbour is the figure of Poseidon astride a monstrous fish, and Mercury appears in a cloud at the top of the view. Around the edges eight heads in clouds, blowing hard, represent the winds. The whole is a masterpiece of the vision and skill of the Italian Renaissance. It has met with better fortune than Rosselli's engravings, and there survive not only some two dozen copies of the view, but also the blocks from which it was printed.

Scholars have long taken an interest in De' Barbari's view, and recently it has been the subject of an important study by Professor Juergen Schulz. His conclusions are of general significance for the development of these detailed, realistic bird's-eye views of towns in fifteenth-century Italy; of particular interest are his answers to the questions why and how the view was drawn. It would be out of keeping with the age that produced it to suppose that it had no more elevated a purpose than to help people to find their way round the city, or simply to show what it looked like in the manner of a travel poster; rather it should be seen as a symbol of the city, epitomizing its size, its wealth and its power. It would be in keeping with views of some other cities in the early sixteenth century if it had been produced to mark a special occasion; thus a woodcut of Cologne was published in 1531 after the election there of Ferdinand of Austria as King of the Romans. The inclusion of the date in the title of De' Barbari's view makes this the more likely, but if so we do not know what occasion it commemorated. On the way it was produced Professor Schulz reaches clear and important conclusions. That it was the product of any sort of measured survey, as earlier writers have assumed, he rules out altogether: accurate measurement of distances would be impractic-

able, even impossible, in the maze of Venice's alleys and canals, and the pattern of the distortions in the view – for distortions there are – makes it unlikely that it was based on a network of angles measured between church towers and other high buildings. Rather he sees it as built up from a sort of mosaic of individual sketches made from these vantage-points, a patchwork of pictures fitted together probably in the framework of existing outline plans of Venice and adjusted to achieve the uniform foreshortening, the single perspective, demanded by the overall bird's-eye view. It was the work of an artist, not of a team of surveyors.

Coming as it does at the very end of the fifteenth century, De' Barbari's view of Venice forms a fitting climax to medieval Italy's tradition of bird's-eye views of cities. It demonstrates more forcibly than any other European example just what could be achieved within the limits of the picture-map, achieved by techniques of the artist that owed nothing to the measurements of the surveyor. At the same time we notice that Professor Schulz suggests that De' Barbari may have made use of existing outline plans of Venice as a starting-point, a base on which to bring together his views of individual neighbourhoods. This raises the question whether there may not have been a further quite separate tradition in mapping the cities of medieval Italy, a tradition of outline plans, from vertical viewpoint, in which measured surveys may have played at least a part. One of the more striking pieces of evidence for this is in fact another map of Venice. Three copies are known; one is in one of the manuscripts of Paolino Veneto's *Magna chronologia* but it is utterly unlike the picture-map of Rome (though not so unlike the map of Jerusalem) that it accompanies there in that its basis is a surprisingly accurate ground-plan of Venice and its nearest islands, showing the Grand Canal and

38, 39 This is only a portion of De' Barbari's woodcut view of Venice, 1500, but it shows the work's almost incredible detail and accuracy. Giving an impressive picture of the size, wealth and activity of the city, it is unquestionably the finest of the bird's-eye views of Italian cities drawn in the new realistic style of the late 15th century.

40   the pattern of smaller waterways. Its general shape and proportions, though far from perfect, are very good. Although the three surviving copies were all drawn in the fourteenth century they are thought to derive from a prototype of the early twelfth century: details of the map point to this, and according to a sixteenth-century inscription on one copy it was drawn for the Doge Ordelaffo Falier, who held office from 1102 to 1118. In the context of medieval topographical mapping, its early date, its accuracy and its sophistication combine to make it a quite remarkable production.

This same sixteenth-century inscription says that the map was made by Hellia Magadizzo, surveyor ('meserador') from Milan. This, of course, may or may not be a true record or tradition. But it must be seen alongside other hints that there may have been measured maps of the towns of medieval Italy. A letter written by the Florentine lawyer Lapo di Castiglionchio to his son Bernardo between 1377 and 1381 gives a circumstantial description of a plan of Florence, now lost:

Again I saw some years ago a document ['carta'] in which Messer Francesco da Barberino, civic judge of Florence, a young man of great ingenuity, had drawn ['figurata'] the whole city of Florence, showing all the walls and their measurements ['e la loro misura'], all the gates and their names, all the streets and squares and their names and all the houses that had gardens so that they could be clearly recognized – and there again every street and every place had its name written on it in his own hand.

Beside these we should probably place some of the representations of towns on the district maps of north Italy – certainly the plan of Verona on the mid-fifteenth-century map from which ill. IV is taken. This shows the city's walls, monuments and street-frontages in elevation and perspective, but it places them on a detailed and substantially accurate outline plan of the city that belongs to quite a different tradition from the contemporary bird's-eye views. Whatever their standards of accuracy or detail, all these town plans are quite distinct from the bird's-eye views both in concept and in technique.

Whether they owe anything to the work of surveyors is more doubtful. Surveyors (*mensuratores*) were employed by medieval Italian cities; their first recorded appearance in Tuscany is at Pisa in 1164. Their work consisted of measuring plots of land, streets and so on for various official purposes. But the only record from medieval Italy that directly links them with these plans of towns is the sixteenth-century note on the Venice map; a connection between maps and surveying is one that could occur naturally to someone in the sixteenth century, so that the annotator might automatically have called the draughtsman a *meserador*, meaning no more than that he was the map-maker. Whether or not medieval Italian surveyors normally recorded their work graphically in the form of plans is something that we may learn when we know more about the local maps and plans of very small areas that were produced in medieval Italy. It is just possible – not more – that these outline plans of towns, like the Naples boundary map mentioned in the last chapter, are evidence of the union in medieval Italy between map-maker and surveyor that certainly did not occur anywhere else in Europe until the sixteenth century. But whether or not they were made by professional surveyors, measurement of a sort must have played a part in drawing up these outline plans, though this measurement may have consisted of no more than simply counting paces along the ground. Castiglionchio's description of the plan of Florence specifically mentions the measurements of the walls, and some form of measurement is implied by the standard of the outline on the Venice and Verona maps as well as on Sanudo's and Paolino's map of Jerusalem, though it is possible, as Professor Schulz has suggested, that the Italian plans are based on official maps that were developed from very crude prototypes by a gradual, centuries-long, process of adjustment and correction, rather than the result of a single careful survey. But whatever their history and techniques – and we have more to learn about both – these town plans are of great significance. For one thing, unlike any other topographical maps from

40 This early-14th-century plan of Venice is probably a copy of one drawn two hundred years earlier. It seems to embody the concept of a ground-plan drawn true to scale; if so, it is the only such plan to survive from medieval Italy.

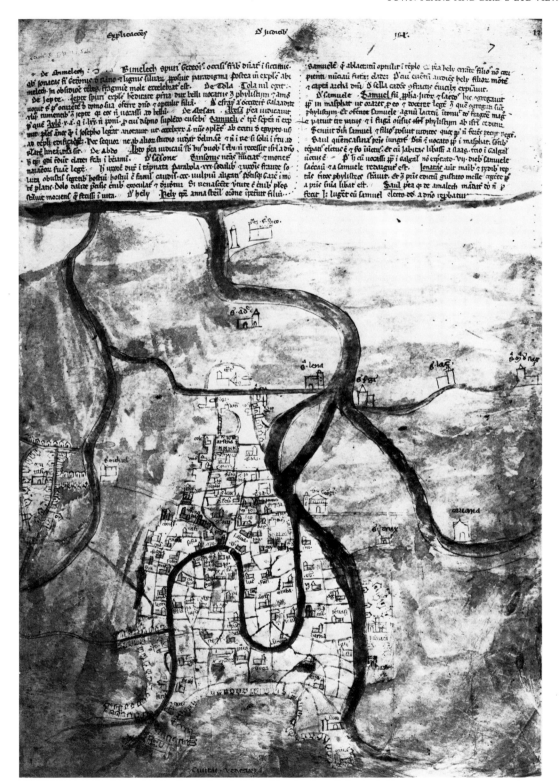

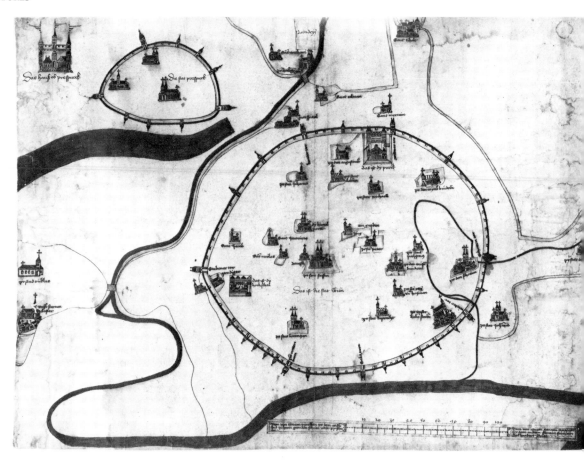

41 Europe's first local map since Roman times to be drawn explicitly to scale, this plan of Vienna was copied in the mid-15th century from one of about 1422. Top left is Bratislava ('Die stat prespurk') with a view of its castle.

medieval Europe, they have clearly attained the concept of uniform scale. For another, just as the north Italian district maps show that Venice was using maps in the work of government before any other state in Renaissance Europe, so the present plans suggest strongly that maps were used in civic administration in northern and central Italy at an earlier date still, something unparalleled elsewhere. All in all this group of Italian town plans is of great importance in the history of topographical mapping.

No less important, for it belongs to the same 41 group, is the earliest known plan of Vienna, a rather later copy of an original of 1421–2.

Besides the main plan it includes, as a sort of inset, a plan of Bratislava (Pressburg). At first sight it resembles one of the bird's-eye views of towns in the older style, for the only details it shows either within the city walls or outside them are the principal churches and other buildings, drawn in elevation and named, but with no street plan or other features. On the other hand the outline of the city walls is drawn with tolerable accuracy and so too, and in some detail, is the pattern of streams in and around the city. But what makes the map really remarkable is a scale-bar at its foot, setting out a scale of paces. It is earlier by some forty or fifty years than any other European map drawn to an explicit scale. That the scale is not quite accurate is beside the point. What is important is that it makes explicit what we could only assume from the Italian plans, that in this one tradition of town plans the concept of a uniform scale had been attained in medieval Europe. That the

Vienna map took its inspiration from Italy there can be little doubt. It may also have owed something to the interest in geography and, perhaps, geographic mapping that we find at this period in the nearby monastery at Klosterneuburg. Certainly it seems to have been entirely a local production – all its inscriptions are in German. But it is only in Italy than any contemporary parallels can be found.

Towards the end of the fifteenth century we find plans and bird's-eye views of other towns outside Italy, but here too it seems most likely that it was from Italy that the idea came, whether directly or indirectly. It is not irrelevant that among the various schools of late-medieval painting it was only in Italy that the bird's-eye view was a normal convention in portraying landscape; and whereas artists from the Netherlands and Germany included views of towns in their paintings, sometimes as seen from a height, these would be of imaginary places, not, as in Italy, of actual towns associated with the artist or his subject. Thus it may well be significant that the only detailed bird's-eye view of a medieval English town comes from Bristol, a port that had regular connections with Italy. It occurs in a chronicle of the city that was written by the town clerk, Robert Ricart, about

1480, and is still in the city's archives. At the centre is a recognizable picture of the town's High Cross, the walls and their four gates are shown less realistically, and the four main quarters of the town are filled with houses drawn conventionally. Compared with the products of Rosselli's workshop it is naïve in concept and crude in execution, but even so it is not impossible that there is a close relationship between them. However, a plan of 1495 showing Rodez in the south of France is connected 42 rather with the outline plans of towns that may or may not be based on actual measurement. Whereas the walls, gates, cathedral and a few other principal features are shown in simple bird's-eye view, most of the houses lining the streets are drawn in a vertiginous mixture of perspectives that recalls, but far surpasses, the plan of Verona on the mid-fifteenth-century IV map of the city and its territory. Here, as on other north Italian district maps, we have at the

42 Late in the 15th century the tradition of bird's-eye views of towns apparently spread from Italy to other parts of Europe. This view of Rodez, in the south of France, with its many picturesque details, was drawn in 1495 for a lawsuit over fairs.

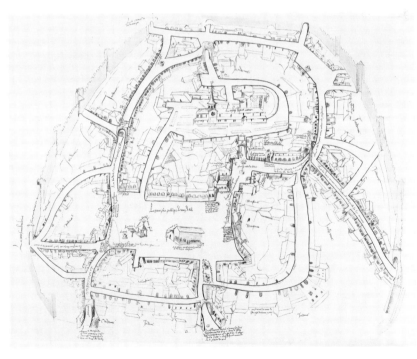

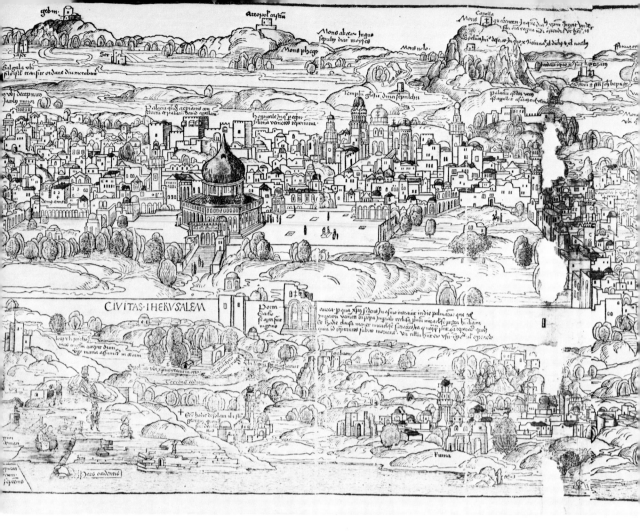

Labels visible on the map image include:

CIVITAS·IHERVSALEM

43 Part of the map of the Holy Land published in 1486 with Bernhard von Breitenbach's account of his travels. On the left is a vastly enlarged bird's-eye view of Jerusalem; on the right the River Nile with Alexandria and Cairo and, in the distant background, the Red Sea.

centre a plan of the city on a vastly larger scale than the rest of the map. On the woodcut map of the Holy Land published in 1486 with Bernhard von Breitenbach's account of his journey there we find the same technique used, but with this difference: that the enlarged representation of Jerusalem at its centre takes the form of a realistic bird's-eye view. That the city view was conceived as separate from the map of the surrounding countryside is underlined, bizarrely, by its being turned through 180

degrees, having west at the top instead of east. The realistic view of Jerusalem is in keeping with the truly pictorial character of the map as a whole; it is a large work, 1 foot broad by some 4 feet long (0·3 by 1·3 metres), and in the illustration we see only a portion, though enough to show its form and general character. Breitenbach's book was the first to be published with woodcut illustrations of this sort; besides the map it contains simple views of places that he visited on his travels.

In its content Breitenbach's map owes a very little to earlier maps of the Holy Land, a great deal to personal observations on the spot. In its form, and particularly in the view of Jerusalem, its greatest debt is clearly to Italian models. Yet it has no obvious direct link with Italy. Breitenbach himself was a canon of the cathedral at Mainz, where his book was first pub-

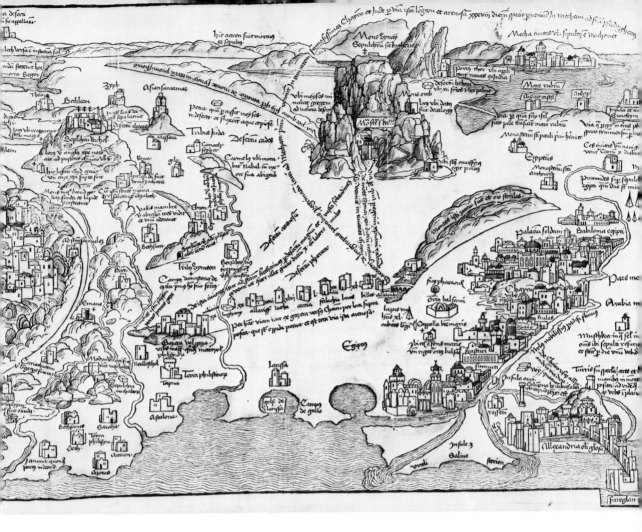

lished; the map and other illustrations were not his own work but were drawn by Erhard Reuwich, an artist from Utrecht whom he took with him on the journey for the express purpose of recording it in drawings. The fact is that by the late fifteenth century the idea of plans or bird's-eye views of towns, having originated in Italy, was spreading more widely. In 1493 there was published at Nuremberg the most elaborately illustrated book yet printed in Europe: the *Liber chronicarum* or *Weltchronik* of Hartmann Schedel, which contained nearly two thousand woodcut pictures. Of these, 116 were of places (other than monasteries) identified by name and these included bird's-eye views as well as ground-level panoramas. But only about a quarter showed their real appearance; the rest were imaginary, a fact that the printer underlined by using the same blocks for different

places – thus the same picture served for Mainz as well as for Lyons, another for Lacedemonia as well as for Milan and so on. Italy had succeeded in spreading abroad her techniques for pictorial representation of towns, but her artists' respect for topographical truth lagged a little way behind. Though not very far: in the sixteenth century, as we shall see, fully detailed town plans and bird's-eye views came to be widely produced, first by artists, then by surveyors. We have seen already how a stylized plan of Jerusalem was placed at the centre of the thirteenth-century Hereford map of the world. About 1510 Heinrich Loriti, a native of Glarus in Switzerland, drew a plan of the universe. At its centre he placed a perspective view of Glarus. The change no doubt reflects an evolving cosmology. But it reflects a developing cartography as well.

83

5

The picture-map
in medieval Europe

WE HAVE seen how in the late fifteenth century the idea of making plans and bird's-eye views of towns seems to have spread from Italy to other parts of Europe, and one might well ask what tradition of topographical mapping already existed in medieval Europe outside Italy. In our present state of knowledge we cannot reach an entirely satisfactory answer. In medieval Italy we found three traditions of topographical maps: the district maps from north Italy, the plans and bird's-eye views of towns, and the local maps and plans of very small areas; and we found that we can only guess that these last really belonged to a substantial tradition of map-making as they have attracted so little attention either from historians of cartography or from writers on Italian local history. In looking at topographical maps from other parts of medieval Europe we meet just the same difficulty. The Italian district maps and town plans and views cannot be paralleled elsewhere; the local maps and plans that we do find may be no more than the merest sketches and because they seem so crude and so trivial they have often been entirely overlooked. We cannot be sure that the pattern of those known to us at all represents the real pattern of those surviving from medieval Europe.

The limitations of our knowledge can be seen from what happened when medieval maps of this sort were investigated in Britain. In 1969, as a first step towards a comprehensive study, a list was published of all local maps and plans made in the British Isles before 1501 that were known to the compilers; it contained fifteen items. By publishing the list and making systematic enquiries among archivists and historians who were likely to have come across such maps the picture was transformed. Of the fifteen original items it turned out that three lay outside the

terms of reference, but additions to the remaining twelve produced a list of thirty maps (or closely linked groups of maps) which have now been published in a collected edition, with detailed studies of each map individually and of the group as a whole. But it is far from certain that all the British material has come to light; indeed, it is almost certain that it has not, for so many of the maps were discovered by the merest chance that there must be others lurking undiscovered. The most that can be hoped is that enough have been found to establish a true pattern of their types, their chronology, their distribution and so on.

It is only in Britain that a comprehensive survey of this sort has been attempted. On the other hand in the Netherlands, unlike anywhere else in Europe, there has been over the last thirty years or so a consistent interest in this kind of map, so that those that have come to light have been published and discussed in journals of local history and elsewhere. Some fifteen medieval local maps from the Low Countries have been noticed in this way, and it seems likely that here too, although others must still be undiscovered, we know enough to survey the group as a whole with some confidence. In France the position is less satisfactory, though the importance of medieval local maps was recognized in the 1960s by an eminent historian of cartography, Dr F. de Dainville. He began a systematic search for them in local archives in south and south-east France with marked success, and in two articles of great interest he published reproductions of some ten maps, which he described as 'a first harvest' in this field, and discussed their significance. Tragically, Dr Dainville died before he could take the work further, and no one as yet has continued it from where he left off. However, he

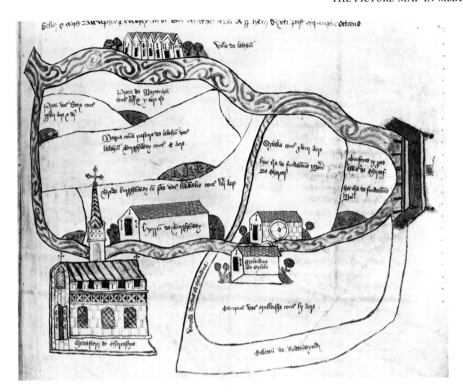

44 Chertsey Abbey, Surrey, with its meadows; at the top the River Thames flows past Laleham, and we also see two watermills and a barn. It was drawn in the mid- or late 15th century in a cartulary of the abbey.

succeeded in showing that, although they have never before been brought to historians' attention, local maps and plans were being drawn in medieval France, and when the harvest is fully gathered it will probably prove considerably larger than that reaped in Britain.

Outside Britain, France and the Low Countries the picture is even more obscure. As we shall see, some medieval local maps have been recorded in south Germany, but in north Germany none. From Poland there survive two sketch maps of about 1464 showing lands belonging to the Teutonic Order and marking the coast-line, rivers and settlements; and there is a record of Polish envoys in 1421 giving Pope Martin V a map (now lost) painted on cloth showing the country's northern provinces. In Spain and Portugal the existence of medieval local maps has never been investigated; but from the early sixteenth century we have a superb manuscript volume, compiled by Duarte de Armas, which gives a plan and ground-level views of each of fifty-seven fortresses along the Portuguese frontier, and this at least strongly suggests that there was already a tradition there of drawing maps and plans of small areas.

The trouble is that the local maps that were made in the middle ages can very easily be overlooked. We have only to look at those from Britain to see how this can happen. A very few are notable artistic productions that could hardly escape the notice of antiquaries. The earliest of all the British maps is a mid-twelfth-century plan of Canterbury Cathedral and its priory, a production of considerable historical interest that has long been well known and has often been reproduced. The same is true of a fifteenth-century map of meadows belonging to 44 Chertsey Abbey: it is brightly coloured and includes a spirited picture of the abbey itself. But many of the others are no more than the merest sketches: a few lines and words put down to illustrate a particular arrangement of properties or to show the bounds of some plot of land. They turn up on blank pages in monastic registers of deeds or attached to statements of rights to property and so on. Anyone working on these documents might reasonably assume that such maps are of no interest apart from any light they throw on the local history of the particular area.

On the other hand we cannot assume, because no medieval local maps have been recorded from a particular country or region, that they are merely waiting to be discovered. They may not exist at all. Careful research and enquiry may have more than doubled the number of these maps known from the British Isles, but all of them come from England: not one has come to light from Wales, Scotland or Ireland. In fact, as we shall see, there are clear signs of regional concentrations, of local traditions of map-making in medieval Europe. One thing that has emerged from analysing the English maps is that if few maps of this sort have been found in Britain it is not just because they have still to be discovered or have simply failed to survive; it is because very few were ever made. This can be demonstrated from the maps found in the muniments of Westminster Abbey and Durham Cathedral Priory, two vast undisturbed medieval archives, each including a large collection of estate records from the thirteenth century onwards. At Westminster two maps have been found; they concern quite different matters and places, and they turned up in different parts of the muniments, but they date from within a very few years of each other in the 1460s and 1470s, and they are both written in the same hand. If maps had often been made in the course of running the abbey's estates it would be a very odd coincidence that just these two should survive. And at Durham we find something very similar: of four maps discovered, three date from the 1440s while the fourth is a late copy of a map originally drawn about the same time, and of the three originals two are in the same hand. In other words, in each of these large and well run monastic estate organizations we have, over some three hundred years, only one or two people who had the idea of drawing sketch maps in connection with problems that arose in the course of business; nor did the idea seem so obviously a good one that it was imitated by their colleagues or successors.

We must in fact accept that drawing and using maps, even of the simplest kind, did not come naturally to people in medieval Europe. We have seen that some primitive peoples thought easily in terms of maps, even to the point of having well developed cartographical traditions of their own, whereas others did not. Medieval Europe shows that this is no less true of more advanced societies. The form a map takes – whether of symbols or of pictures – reflects the general level of a society's culture; the fact of its having any topographical maps at all does not. Medieval people simply did not think of drawing maps for the innumerable purposes for which we are apt to take them for granted. Thus, for instance, the complexities of the ownership of arable strips in open-field farming would be set out in elaborate written descriptions; directions for a journey would take the form of a list of the places to be passed through. To us this seems cumbersome and inefficient; nothing shows more plainly how strange, how alien, the idea of a map was to medieval people. Dr Dainville has described this as something very revealing of their mental attitude, and we have to recognize that it does indeed point to a habit of mind quite different from our own.

In looking at the symbol-maps of primitive societies we found distinct traditions of map-making: small pockets of map-consciousness, so to speak. In medieval Europe we find just the same thing. We have already seen one local tradition: the district maps from north Italy that have no parallel elsewhere in the peninsula. Another example is provided by the local maps from the Low Countries: every one of those so far recorded comes from a long coastal strip from Ostend to Haarlem, stretching inland no further than some thirty miles to a line drawn from Ghent in the south to Hilversum in the north. In Britain we find no similar concentration; the maps may be confined to England, but they come from all parts of the country, from Durham to south Devon. On the other hand there may be a significant grouping of maps in eastern England around the Wash: seven of the thirty known English maps are believed to have been drawn in this area, and if we exclude the earliest of them (from Kirkstead Abbey in Lincolnshire) as a possible chance intruder, we can say more precisely that they all come from within twenty miles of Wisbech. It is interesting that this lies just on the other side of the North Sea from the

45, 46

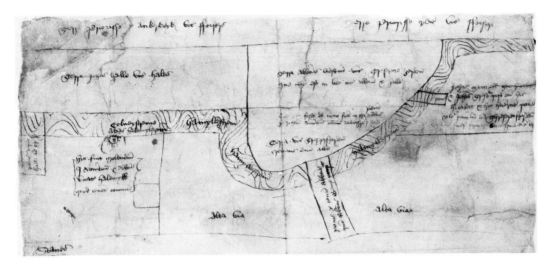

45, 46 The only two medieval maps in the archives of Westminster Abbey were both drawn by the same hand in the 1460s or 1470s. Both are diagram-maps: only ground-level features are shown, though there is a pictorial element in the wavy pattern marking small streams. The map above shows lands at Staines on either side of a stream that runs into the Thames. Below is a more elaborate plan of fishing rights and other properties along the River Colne, the broad band across the middle of the map; the lands at the top are in Harefield, those below in Denham, and the diagonal bars in the river mark a weir.

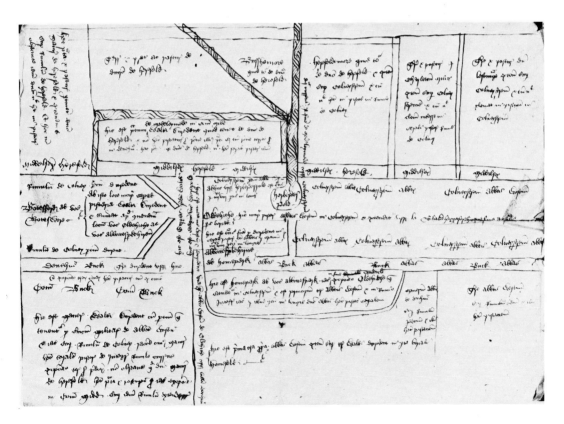

area of map-making in the Netherlands, and interesting too that its landscape would create just the same problems of imbanking, drainage and pasturing that are particularly in evidence on the maps from the Low Countries. But if this was what produced map-consciousness in this part of England it did not limit its application, for only three of the seven maps concern matters of this sort. From France we are not yet in a position to distinguish any particular pattern in the distribution of medieval local maps, but when we have a broader view of the material we may well find some similar grouping there. Certainly in south-west Germany there are signs of what may well be another significant grouping: a sketch map of 1441 showing a monastic estate that straddled the Rhine at Wantzenau, north of Strasbourg, a plan of Ulm about 1480, two picture-maps from Beringsweiler near Heilbronn tentatively dated 56 about 1500, the magnificent engraved bird's-eye view of Lake Constance showing incidents in the war of 1499 – all these may point to another map-conscious area in the mid- and late fifteenth century. This would be particularly interesting because of the important early scale-maps drawn from measured survey that came from this part of 85 Europe: the maps of Erhard Etzlaub of Nurem-87, 88 berg and Konrad Türst's manuscript map of Switzerland in the 1490s, the maps of the Black Forest and Alsace by Martin Waldseemüller in 1507 and those published with his edition of Ptolemy's maps in 1513. The particularly detailed representation of Reichenau on the mid-thirteenth-century Ebstorf world map (destroyed in the Second World War) may point to an even older tradition of local maps in this area. We have much more to learn about these local concentrations of maps, but it is at least clear that topographical mapping did not occur uniformly throughout medieval Europe, and that its geographical distribution may throw light both on its origins and on its influence on later cartography.

So too may its chronology. Of the English maps, the Canterbury Cathedral plan was drawn probably in the 1150s, two others belong to the thirteenth century, and the other twenty-seven all date from the mid-fourteenth century onwards, not in a crescendo but at a steady level of production: from eight to ten in each half-century. The earliest map from the Low 47 Countries is of an area near Sluis in 1307, and

another, from the archives of St Peter's Abbey at Ghent, has been dated 1358, but the rest belong to the fifteenth century and may all be later than 1450. In France the pattern is similar. There the oldest map so far found was drawn in 1357 in a register of Paris University; it shows the line of the River Maas, held to be the boundary between two of the university's *nations* – the groups into which its students were divided, corresponding to their place of origin. None of the maps that Dr Dainville discovered is older than 1422, but there are references a few years earlier to maps that are now lost. In 1395 Jehan Boutillier wrote in his treatise called *La somme rurale* that in putting cases before the Parlement use might be made of 'figure et pourtraict'; this must mean a map. No local map recorded from Germany is older than that of 1441 from Wantzenau on the Rhine. We see how closely this pattern corresponds to what we find in Italy: there the outline plans and bird's-eye views of towns start to appear in the early twelfth century, and the north Italian district maps in the late thirteenth, but it was in the fifteenth century that most of the surviving maps were drawn. At first sight it might seem that this is simply a pattern of survival: that the fifteenth century did not necessarily produce more maps than earlier periods, but simply produced more that have happened to survive. This is unlikely; from England, for instance, records of estate administration (and it is among these that many of the maps are found) start to survive in profusion in the mid- or late thirteenth century, a pattern quite different from the maps. Almost certainly the survival of significantly more local maps from the late fourteenth century onwards means that significantly more were being produced.

So, although there are great gaps in our knowledge of the local maps from medieval Europe, what we do know of their chronology and distribution makes a consistent pattern. Until the fourteenth century topographical maps were almost unknown. Then, first in Italy and next probably in England, people became aware of the possibility of making maps for one reason or another, so that in the fifteenth century we find them appearing widely in western Europe. But even then one could not say that the idea of making maps had really caught on: although in a few

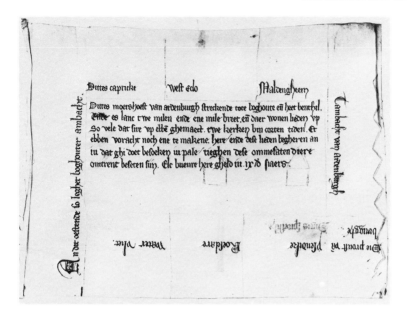

47 The oldest map known from the Low Countries, drawn in 1307 to show an area near Sluis. The outline gable-ends of buildings are two churches. The map is on a roll of payments; the right-hand edge is indented.

areas there seems to have been a genuine local tradition of map-making it is quite possible that many fifteenth-century maps were drawn by people with no previous knowledge of maps at all, who saw themselves as doing something entirely new, using a quite original method to set out landscape on paper or parchment. The question we now have to ask is how this pattern can be explained. We have seen how in certain societies a tradition of symbol-maps gave way, at a particular stage of cultural development, to a tradition of picture-maps; but here in late-medieval Europe what we have is the apparently spontaneous growth of picture-maps without the precedent of symbol-maps – indeed symbol-maps are mostly the products of less advanced cultures, and if early medieval Europe had drawn topographical maps these would probably have been picture-maps of the same sort as eventually appeared.

There are at least three ways by which medieval Europeans outside Italy may have been brought to the idea of drawing maps. One originated in the various kinds of bracketing,

tabulation and diagram that scholars and scribes were apt to use in manuscripts when setting out relationships of any kind, from theological propositions to the parts of an estate account. It needed only a very slight mental jump to apply the same sort of system to topographical relationships. Among the Italian maps discussed in the last chapter, the earliest map of Rome, deriving from a twelfth-century original, is closely related to non-topographical diagrams; indeed our surviving copy occurs in a manuscript where it is one of a whole series of diagram-like illustrations. The earliest map from the Low Countries, the plan of 1307 of an area near Sluis, 47 is scarcely more than a systematized layout of wording. Panels in a double frame give the names of settlements in their relative positions, while two outlines of gable-ends of buildings represent the two churches built to serve the area which lay between the parishes of Aardenburg and Boechoute; the writing in the central portion, telling of the growth of population that had made the two churches necessary and now called for a third, so dominates the map that one can view it practically as a written document in schematic form. Similar in concept, though with much less wording, is the map of Wildmore Fen in Lincolnshire that was drawn up between 1224 and 1249 and was copied into a psalter at Kirkstead Abbey; along the bottom a wavy line marks the River Witham, a diagonal across the page represents the boundary between the baronies of Bolingbroke and Scrivelsby, and on either side names appropriately placed mark the properties that belonged to Kirkstead and other manorial lords. Particularly interesting in this connection is the other thirteenth-century map from England. It shows the layout of springs, pipes and cisterns at Wormley in Hertfordshire from which Waltham Abbey, three miles away, took its water supply; it was probably drawn in the 1220s, though the copy we have is slightly later. Fifteen miles from Waltham was St Albans Abbey, and there, just at that time, was living Matthew Paris; he was not only the author of the maps of Britain and the Holy Land and the itinerary to southern Italy, but was also the copyist or illustrator of various works containing diagrams, among them windroses, astronomical diagrams and diagrams for telling fortunes. In its general style the Waltham map is remarkably similar. Matthew Paris is known to have had connections with Waltham and, directly or indirectly, it may well have been he who suggested the idea of setting out the arrangement of pipes and so on at Wormley in diagrammatic – or cartographic – form.

These are all early examples. But there is no difficulty in finding fifteenth-century maps too that do little more than set out information diagrammatically with the minimum of graphic or topographical elements. Good examples are the two maps from Westminster Abbey. One is concerned with a claim to some land at Staines (now in Surrey), the other with fishing rights in the River Colne between Harefield (now Greater London) and Denham in Buckinghamshire. On both there is a lot of wording, and although its arrangement corresponds to the topography of the areas and lines mark the rivers and boundaries – there is no question but that they can be seen as maps – the graphic elements play a subordinate part in their composition. The same is true of a plan of 1487 from the archives of the Great Council (*Groote Raad*) of Mechelen: it shows watercourses on the island of Overflakkee at the mouth of the River Maas and consists of a few lines and a lot of words, though it is enlivened by a vigorous if crude sketch of a windmill. The 1441 map of the estate at Wantzenau is another map of this sort: a schematic arrangement of rectangles, with wording in each, bisected by a series of wavy

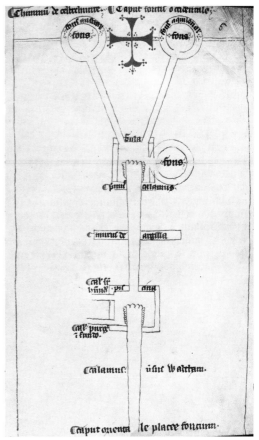

48 The source of Waltham Abbey's water supply at Wormley: a mid-13th-century copy of a plan of the 1220s. Circles mark the springs, rectangles tanks, scalloped lines the perforated ends of pipes. The cross at the top is the earliest direction pointer known on any map; its splayed foot marks the east.

lines representing the Rhine. A very interesting case of a diagram that teeters on the verge of being a map comes from Gloucester in 1455. It is a rental, a list of the town's householders, and like many such lists it proceeds street by street, naming the occupants in the order of their houses; what is less usual is that they are listed in double columns, corresponding to the two sides of each street, while tiny pictures of the town's principal landmarks – the town crosses, the churches, the pillory and others – have been sketched in at the appropriate places in the lists. Nothing is drawn in plan, none of the sketches even takes the form of a bird's-eye view; it is not a map. But it shows just how easy it is to advance

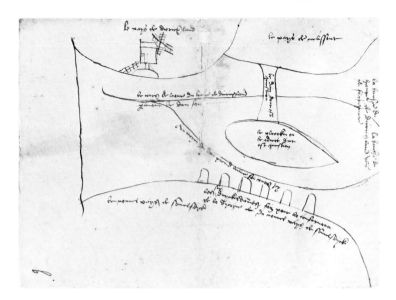

49 Many medieval local maps originated in lawsuits. This plan of 1487, showing part of Overflakkee at the mouth of the Maas, was drawn in a dispute between two neighbouring landowners over waterways and embankments.

50 Roll of 1455 listing all the houses in Gloucester. Pictures of the town's principal landmarks appear at the appropriate points; here we see Christ Church and the Franciscans' church top right and the church of the Dominicans below.

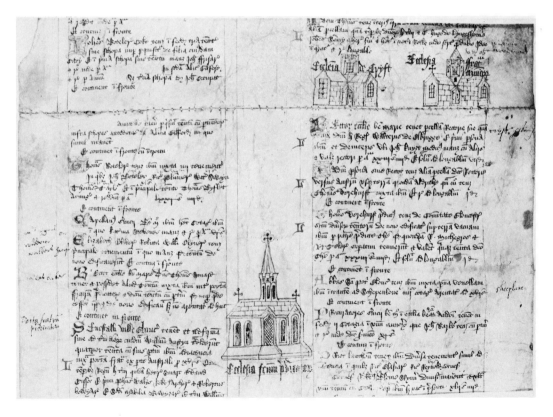

to the map from a formalized layout of wording with a topographical interest.

The non-topographical or non-cartographic diagram was thus one route by which north-west Europe arrived at the topographical map. Another was discovered by Dr Dainville in his work on the maps in French archives. Several, he found, were referred to in contemporary notes by the curious name *tiberiad*. This harks back to the treatise written in 1355 by Bartolo da Sassoferrato, entitled *De fluminibus seu tiberiadis* ('On rivers or tiberiads'), in which he discussed the sort of legal problems that arise over rights in rivers and streams, using diagrams to show how they could be resolved. *Tiberiad* refers to the River Tiber: Da Sassofer-

present we know so little about the local maps produced there. We have, of course, only the word *tiberiad* to link the French maps with Da Sassoferrato's treatise and to suggest that the idea of making them came from him, but the date of their appearance in the early fifteenth century fits well and certainly there can be no question that in both France and the Low Countries local maps were made for production in law-courts, either to show what one or other party was claiming or to serve as a basis for agreement. Thus the plan of Rodez in 1495 was [42] drawn for a suit over fairs, and a few years earlier, in 1491, a map of Avignon and its surroundings had been used in a complaint to the Pope about infringements of rights there by royal officers from Languedoc and Provence. We find just the same in the Netherlands. There the word *tiberiad* has not been recorded, but several of the surviving maps were demonstrably made to be shown in court. One of the lower reaches of the River Scheldt, dating in [51] its original form from 1468, was made on the order of a committee that was collecting evidence from both parties to a suit; it was altered later for use in other cases. The sketch map of 1487, already mentioned, showing [49] watercourses on Overflakkee, comes from the files of cases heard before the Mechelen Great Council. There can be no serious doubt that the formal demands of litigation brought about the production of some, probably many, of the medieval local maps from France and the Low Countries. It is very likely that this practice originated in the law-courts – or at least the law schools – of Italy, and that this was one way that the idea of drawing local maps was introduced to these countries, where, of course, once introduced, it could easily be applied to other purposes as well.

On the other hand we cannot explain the appearance of local maps in medieval England in this way. English common law, however much it may have owed to the civil law of the

rato wrote the work when he was staying in a house by the river, which brought these questions to mind, and he was helped in the diagrams by his friend Guido da Perugia, whom he called his mentor in surveying ('meus . . . in geometria magister'). Whether plans of this sort were drawn up in medieval Italy to place before courts of law we simply cannot tell, given that at

Continent, was following an independent line of development. Da Sassoferrato's treatise may well have been unknown in England; no copies have yet been traced in English manuscripts. There is nothing to suggest that the medieval English law-courts ever saw, let alone asked for, a map or plan in the course of hearing a case, and certainly not one of the surviving English maps can be shown to have been made for production in court. The only possible exception dates from the very end of the fifteenth century, 1499. This is a formal document setting out the history of a property in St Mary Arches Lane in Exeter, and at the top is a well drawn plan, in ink and wash, showing the street and boundaries with elevations of the buildings and walls of the frontages. Interestingly, this document was drawn up by a public notary, whose work lay in the tradition of the Continental civil law. But it is clear that we must rule out the treatise of Da Sassoferrato and the demands of law-courts as agents in introducing local maps to medieval England.

If so, how can we explain the sudden, if limited, appearance of local maps in England in the mid-fourteenth century? One possibility is that they came by way of the building-plan, and this may indeed have contributed to the introduction of local maps elsewhere in Europe too. But whereas we have something approaching tangible evidence that maps might originate in non-topographical diagrams and in the demands of law-courts we have nothing to show that building-plans inspired the production of local maps beyond a mere coincidence of dates, and even this is partly conjectural. Ground-plans (*plats*) were drawn as a normal part of constructing any substantial building in late-medieval England; they were one of the techniques of the architect's or mason's craft. We know this from many references in building-contracts, although only one medieval English building-plan survives; this shows part of a court, with doorway and staircases, that was built at Winchester College about 1390. From the Continent, however, we have many more. A particularly notable collection, including elevations as well, is at Vienna in the archives of St Stephen's Cathedral; they come not only from Vienna and nearby towns, but from Ulm, Prague, Strasbourg and Cologne, having been brought to Vienna in the normal course of business by itinerant masons. The oldest elevation in the Vienna collection is not earlier than 1350, and the oldest plan is probably of the early fifteenth century. In England the first known reference to a building-plan is in 1380. It is at least arguable that plans on parchment or paper first became a normal part of building techniques in the mid-fourteenth century. Although there are building-plans in the notebook of the thirteenth-century French architect Villard de Honnecourt this need not mean that plans were already in everyday use by builders: lines scratched on the ground or on tracing-floors were probably the precursors of drawn plans, and these may still have been the normal method of planning a building. But when drawn plans began to be used by builders – and they would be shown to their clients, as the building-contracts make quite clear – it is quite possible that a few nimble-minded persons realized that the same technique of the plan or map could be applied to many other problems of business or administration as well. Some such process as this could explain the chronology of local maps in medieval England. But this is only conjecture, and a different explanation may some day be found for their sudden appearance in the mid-fourteenth century.

At present, however, we can see three routes by which medieval man outside Italy may have arrived at the idea of the topographical map – besides, of course, the inventive genius of isolated individuals, which should certainly not be overlooked. There was the progression from non-topographical diagram to map; there was the introduction from Italy of the use of maps in law-courts; there was possibly also the extension to other purposes of the concepts embodied in the building-plan. But however the idea was arrived at, to what purposes was it applied? What did people draw local maps for in medieval Europe?

One purpose, of course, we have already seen: for production in law-courts. But it was not only this that led to drawing maps in connection with disputes and litigation. When there was a disagreement over landed rights – property bounds, pasturage, fishing rights and so on – one or other party might well illustrate his claim in the form of a map. This would not be to show to a court, and indeed litigation might

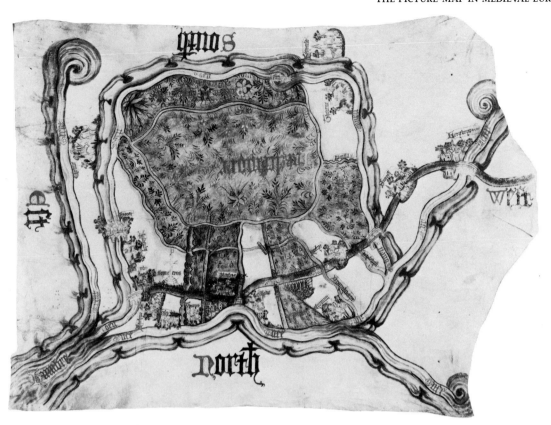

52 Some picture-maps give us a real glimpse of the medieval landscape. On this later 15th-century copy of a map of Inclesmoor first drawn in 1405–8, we see villages, bridges, wayside crosses, fenced enclosures, trees and flowers. Its bounds are the rivers Trent (left), Ouse (bottom) and Aire (right).

not even be in question, but it would set out clearly what was at issue, might perhaps serve as a basis for discussion with the other party and, above all, would be a permanent record in the estate's muniments to show future generations just what was claimed. Consequently some of the English maps are found in cartularies, collections of an estate's deeds and other records copied into registers for easy reference, as at Durham Cathedral Priory and the abbeys of Peterborough and Chertsey. Very few of the surviving maps are known to have been taken into court, but a great many can be connected with specific claims and disputes: about a third of all those from England, and a substantial majority of those from France and the Low Countries. These are not confined to any particular part of the period. Kirkstead Abbey's map of Wildmore Fen, dating from between 1224 and 1249 and already mentioned as a diagram-derived map, is closely connected with a whole series of disputes over the abbey's rights of pasture there. And nearly two cen-

turies later a map was drawn for the Duchy of Lancaster when in 1405–8 its officials were at work collecting evidence for a suit against St Mary's Abbey, York, over rights of pasture and peat-cutting in Inclesmoor, an area in Yorkshire (now Lincolnshire) near Goole. While covering the whole of the moor and its surrounds, the map goes into particular detail in the area of the dispute, giving names and other information taken from relevant records of a century before. It seems likely that the two versions of the map that we have both derive from a now lost original. One was bound into the Inclesmoor section of one of the Duchy's cartularies, probably soon after 1416, while the other, a 52

53 An account of rights over the pool of Scamandre in the Carmargue, illustrated with a picture-map. The original map, which still survives, dates from 1479; what we see here is a close copy made in the 17th century.

highly coloured and very ornate map, dates probably from the later fifteenth century; in both cases we see the map, drawn up for a specific dispute, being used for permanent reference or record. Two examples from Ghent show how maps might be used to set out informally what had been agreed in settlement; both concern tithes and other rights belonging to St Peter's Abbey, illustrating agreements of 1358 over the parishes of IJzendijke and Oostburg, and of 1480 over the Braakman area. From the Carmargue in the south of France the abbot of Saint-Gilles' claim to jurisdiction over the pool of Scamandre is set out in a document 53 of 1479 which is illustrated by a sketch map. And, moving from the very local to the national level, we find that in 1444 Philip the Good,

Duke of Burgundy, disputing boundaries with King Charles VII of France, had maps made by his physician and councillor Henri Arnault, also known as Zwolle from his home-town in the Netherlands; we know this from Zwolle's receipt for payment he had received.

for having made . . . plans (figures), showing the distances between places and the bounds of certain regions and marches of the Duchy and County of Burgundy . . . so as to see clearly the towns and villages that are included in the Duchy, and also those that belong to the kingdom; which plans have been sent to my said lord in his lands of Flanders, so as to avoid and guard against the encroachments that are made every day by the people and officials of the king.

We have two maps of 1460 that relate to later disputes over the Duchy boundary; one of them shows three villages, the lands between them 54 and, in the distance, the River Saône.

But even where there was no particular dispute to illustrate or contention to prove,

96

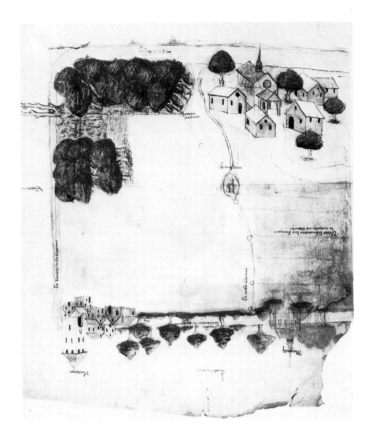

54 The boundary of the Duchy of Burgundy near the villages of Talmay, Maxilly and Heuilley, 1460. The picture-map is drawn with opposing horizons. In the distance, beyond Heuilley, we see the River Saône.

maps might be drawn simply as guides to properties of one sort or another. The two earliest English maps – of Canterbury Cathedral in the mid-twelfth century, of Waltham Abbey's plot of land at Wormley in the early thirteenth – both show water systems that included buried pipes. A plan seems so obvious a solution to the problems these present that it is surprising that we have only one later example; this is a mid-fifteenth-century parchment roll, some 10 feet (3 metres) long, showing the conduits bringing water to the Charterhouse in London from springs at Islington, two miles away, and it is particularly interesting as it includes a plan of the Charterhouse itself, showing individual cells and other rooms. Property of quite a different sort is shown on a plan of about 1420 in a cartulary of St John's Hospital at Exeter: on a page that sets out the rather involved succession of rights in a piece of land used for drying cloth after fulling it gives a tiny sketch plan showing the arrangement of the tenter-frames on which the cloth was

stretched. On quite a different scale of elaboration is another map from Devonshire, now in the Royal Albert Memorial Museum in Exeter. It shows the whole of Dartmoor as a circle, with rivers issuing from it and with pictures of Tavistock, Buckland and other places around it; on its south side Harford and Ugborough Moors appear in greater detail and on a far larger scale. It seems likely that it was drawn, at the end of the fifteenth century or a little later, not in connection with a particular claim but simply to illustrate the pasture rights of the manorial lords of Ugborough, including the so-called venville rights over all Dartmoor.

Outside England we know of no maps drawn simply to illustrate property rights without reference to particular claims or disputes, but it seems likely that examples will eventually turn up in France, the Netherlands and elsewhere. The same is true of another group of local maps from England: the five that were drawn to illustrate historical writings. Robert Ricart's bird's-eye view of Bristol is one of these;

although it clearly shows the late-fifteenth-century city, its position in his chronicle suggests that it is meant to represent Bristol at the time of its legendary foundation. A map of the area around the village of Barholm in south Lincolnshire illustrates a late-fifteenth-century manuscript which may well be the earliest of English parish histories and which far outstrips even the most imaginative of its successors in its bizarre mixture of fact, anachronism and legend. The much more reputable early-fifteenth-century chronicle of Thomas of Elmham, monk of St Augustine's Abbey at Canterbury, contains a plan of the altars in the abbey church and a map of the Isle of Thanet that illustrates the 'run of the deer': the strangely contorted boundary of the abbey's manor of Minster, traditionally the course followed by the pet deer of Queen Domneva of Mercia, who was given as much land as it encompassed at a single run. Finally there is the map of the village and fields of Boarstall in Buckinghamshire, probably intended as the frontispiece of the cartulary that its manorial lord, Edmund Rede, had drawn up in the 1440s; he was a man with a strong interest in his family and its history, and at the foot of the map we see a legendary incident in which his ancestor, Nigel the Forester, presented to King Edward the Confessor the head of a ferocious wild boar that he had just killed, receiving as a reward the hereditary wardenship of Bernwood Forest. These antiquarian maps are particularly intriguing. For one thing they foreshadow the interest taken in maps by later English antiquaries – men such as John Leland and Lawrence Nowell in the sixteenth century, John Speed and William Dugdale in the seventeenth – many of whom drew maps themselves. For another we find a similar connection between antiquaries and early maps in Italy: we have already seen how Flavio Biondo's book on the antiquities of Rome lay behind a fifteenth-century bird's-eye view of the city, and we find a similar emphasis on historical information in a number of the maps drawn in sixteenth-century Italy.

Another type of map recorded in both France and south Germany was that intended for display, the map meant to be on general view and not just to be consulted on particular occasions. None is known from the Low Countries; from England we have a reference to a *mappa mundi* that was among the mural decorations painted for King Henry III at Westminster in the 1230s, but at this early date this almost certainly meant literally a map of the world, not just any map at all. In Italy we saw how, at Florence, Siena and elsewhere, maps or bird's-eye views might form part of mural decorations, and how too, at the end of the fifteenth century, large realistic bird's-eye views of cities were being printed, having an underlying significance that is not entirely clear to us, but certainly with the intention of being put on display. From fifteenth-century France, although we have no surviving maps of this sort, we have records that show that they once existed. In 1440 the Duke of Orleans was given a cloth, long enough to go round a cloister, showing the course of the River Loire with its towns and bridges, while an account-book of 1472 mentions among properties in the castle at Angers three cloths painted with representations of towns. One of these last, showing Tours, is called a *mappemonde* and was thus probably a plan or bird's-eye view, and the others too may well have been picture-maps, not ground-level panoramas. From Germany we have what was the first printed map of this kind to be produced outside Italy. It shows the whole of Lake Constance, with a westward extension as far as Basle and Berne, in the style of a bird's-eye view. Villages and towns are named, but they contain imaginary buildings, not drawn from life. Shown in the landscape are incidents from the war of 1499 between the Emperor Maximilian I and the Swiss Confederation, as seen (politically) from the Swiss point of view. Its artist is unknown: his monogram 'P W' or 'P P W' is engraved on the map, but he has not been conclusively identified. The map was engraved on six copper plates; its size overall – about $1\frac{1}{2}$ by $3\frac{1}{2}$ feet (0·5 by 1·1 metres) – makes it about the same size as Rosselli's view of Florence, and certainly in the scale of its vision and its sheer panache it deserves to be placed alongside the work of Rosselli and De' Barbari.

55 Thanet in the early 15th century from a chronicle, by Thomas of Elmham, of St Augustine's Abbey, Canterbury. East is at the top; bottom left is Reculver church. Thomas's chronicle also included a plan of the altars in the abbey church and facsimiles of charters.

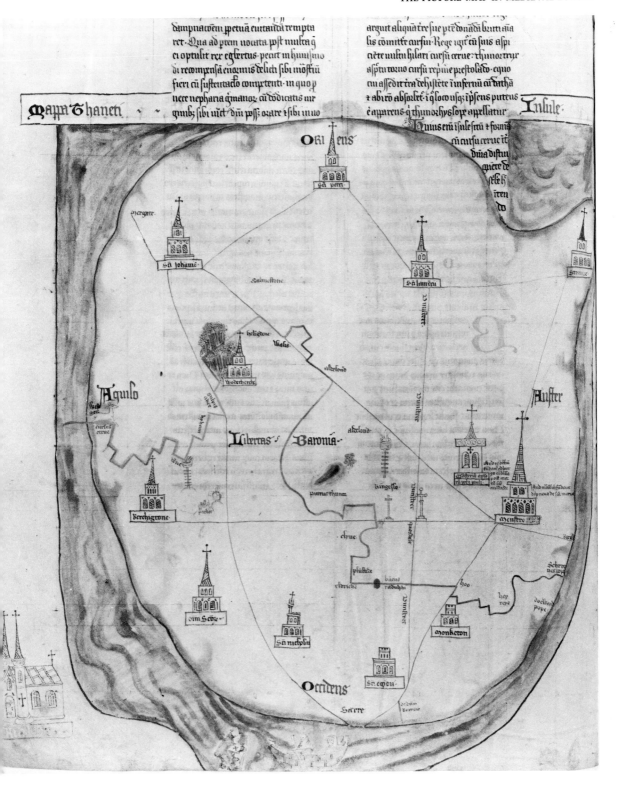

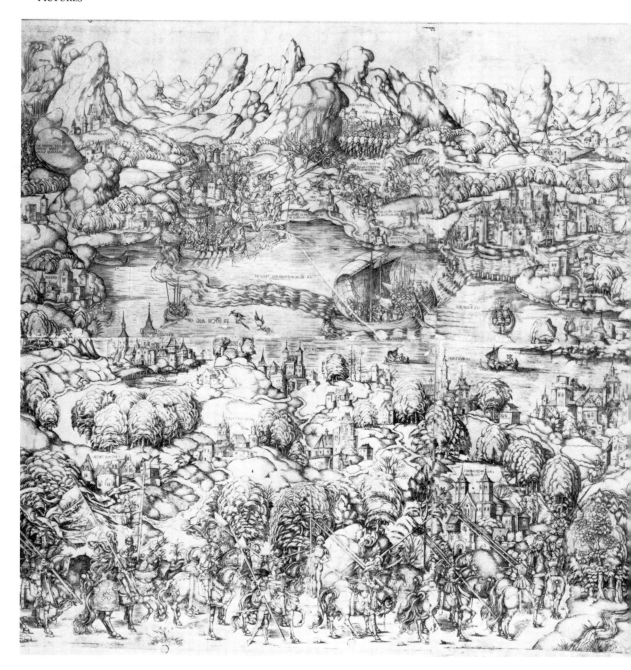

56 Picture-map of Lake Constance and areas to the west, showing scenes from the war of 1499: a woodcut printed from six blocks, probably about 1505. South is at the top; the Alps close the horizon. It is interesting to compare this with the map of 1578 (ill. 1), which partly covers the same area. Both maps consist of a multiplicity of tiny pictures, each drawn to a separate horizon, but the later map is more advanced in concept, for it sets them in a framework that is true to scale throughout.

One purpose for which we might expect local maps to be drawn in medieval Europe, one that to our way of thinking a map is peculiarly fitted for, is to embody the results of an estate survey: to set out cartographically, instead of in written form, the minutely detailed descriptions of landed property – a group of fields, a single manor or an entire estate – that were so often made to meet the needs of medieval royal or private administration. From northern France the elaborate written survey made for Abbot Irminon of Saint-Germain-des-Prés dates from the early ninth century. In England we find that already in the Domesday Book of 1086, the masterpiece of medieval royal surveys, actual measurements are given of a few meadows, woodlands and vineyards, and from the thirteenth century onwards it became quite usual to include in surveys the carefully measured size of each plot of land. Also in the later middle ages the form of survey known as a terrier became common: a systematic account of every strip or other piece of land in a particular field or village, giving the bounds and measurements of each. Often the description is so full that one could perfectly well use it to construct a detailed map; these surveys are in fact almost written maps, what Dr Dainville called *cartes parlantes*. But here, where a map would be so much quicker, simpler and more effective than a written description, the tiny conceptual jump that was needed seems never to have been made. Only twice have medieval local maps so far been found associated with estate surveys, in each 57 case from England. From Shouldham, in Norfolk, we have two very rough sketches of groups of strips in open fields accompanying a terrier of 1440–41; from Tanworth in Arden, in Warwickshire, a group of small fields appears on plans connected with a survey made at the very end of the fifteenth century or a little later. In neither case are the plans part of the final survey; they are simply among the rough notes and drafts. And although we might wonder whether the drawing of sketch maps in the fields was not a normal part of drafting surveys of this sort – a technique of the surveyor, just as the building-plan was a technique of the late-medieval builder – it seems unlikely; the Shouldham sketches in particular suggest that the draughtsman was unused to drawing maps or plans of any sort. The value of maps for this

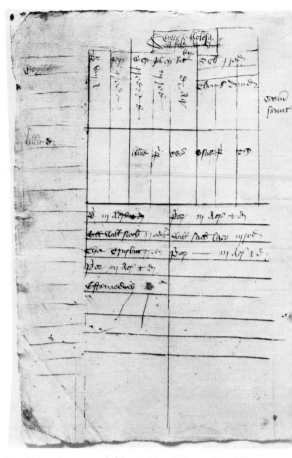

57 Strips in open fields at Shouldham, Norfolk, 1441; on some are written their size and owner's name. Surprisingly, the medieval surveyor seems hardly ever to have used such sketches to help him record fields and other holdings.

purpose – to us so obvious – seems to have eluded our medieval forebears.

Apart from these two sets of plans from England, which do not provide very convincing evidence, there is nothing to connect our local plans from north-west Europe with surveyors, the men who measured lands and drew up written surveys. In England even by the fifteenth century we find no class of professional surveyors, though already two centuries earlier there are signs that, as one might expect, surveys would be made by those estate officials or others who had particular experience or skill. In the Low Countries, however, we find men who are clearly professional surveyors. In 1281 the city of Bruges made a payment to someone called Johannes Landmetra; the name strongly suggests that surveying was his calling or principal occupation. And at the beginning of the fifteenth century Jan Matthijssen devoted a section of his law-book for the town of Brielle (at the mouth of the River Waal) to the work of the surveyor, describing his qualifications, his apparatus, his duties and the oath he must swear, before starting a survey, to measure correctly and record the results honestly. But Matthijssen says nothing about making sketches, plans or maps, either as part of the process of surveying or as its end-product, and though we do not know who drew the local maps surviving from the medieval Netherlands there is nothing to suggest that they were surveyors; certainly, as we shall see, it was the professional artist rather than the surveyor who would be called on if a map were needed in the Low Countries in the early sixteenth century. This is what we find in fifteenth-century France. In 1423 Antoine Actuhier, ordered by the Dauphin to have a map made of the counties of Valentinois and Diois, employed a painter, Jean d'Ecosse, to undertake the work. He spent twelve days touring the area with a guide, and they took with them a tent, a portable table, a box to hold the colours, three paper sketch-books and some green cloth on which the map was to be prepared – equipment for an artist, not a surveyor.

Certainly precise measurement, which lay at the root of the surveyor's work as much in the middle ages as it did later, played no part in the construction of medieval local maps outside Italy, with the single exception of the mid-fifteenth-century plan of Vienna. Not one of those known from England, the Low Countries, France or Germany seems to make the slightest attempt at consistency of scale, even when measurements are actually noted on the map. This is the case in plans drawn in the 1470s of four building-plots belonging to the estates assigned to the upkeep of London Bridge: the ground measurements are given in some detail, but the lines on the map are not drawn proportionately. The same, incidentally, is true of the solitary English building-plan from Winchester College, but even this need not surprise us: if the outline was roughly the right shape and measurements were duly noted, the plan would serve its purpose without being drawn to scale. This, then, is one thing that all maps considered in this chapter have in common: however diverse they may be in date, origin, style or purpose they are all alike in paying no regard to scale. And another is that any features marked on the maps are shown pictorially. Sometimes, of course, the map consists only of an outline, such as the plan of tenter-frames at Exeter or the plan of 1358 from St Peter's Abbey at Ghent. But if anything more than this appears its form is pictorial, from the elaborate elevations of Canterbury Cathedral and its precincts on the earliest map from England to the simple gable-end outlines that mark the two churches on the earliest map from the Low Countries. These two characteristics – the lack of scale and the strong pictorial element – place the local maps of medieval Europe firmly in the category of picture-maps.

6

The picture-map in the Far East

To PASS from the cartography of medieval Europe to that of ancient China is to discover a surprising contrast. Europe in the middle ages was unfamiliar with maps and plans. We have seen how the idea of using them was practically unknown before the thirteenth century, and even then developed mostly in traditions restricted to a particular region or to a particular profession or craft. In China, on the other hand, there are many references to maps in writings from the third century BC onwards, and officials, soldiers and scholars all seem to have been perfectly familiar with their use. Given that in China the word for map (*thu*) is ambiguous, as it is in other languages, and that we have learned to be wary of scholars who take for granted that past ages were as familiar with maps as we are ourselves, we might wonder how far this should be believed and whether in the earliest period at least these references to maps may not arise simply from mistakes in interpretation. But the discovery in 1973 of Chinese topographical maps actually dating from the second century BC sets any doubts at rest.

The maps were found near Chang-sha in south-east China, in the tomb of a man who must have been an important military commander and who was buried in 168 BC. They are drawn on silk and were in a box with other silk manuscripts; other objects in the tomb included paintings and weapons. The maps had been folded many times, and as they had disintegrated along the folds a great deal had to be done to conserve and restore them. One of the maps is a plan of a town, an administrative centre, and the other two are district maps; one shows a wide region, some three hundred miles square, and the other covers a smaller district in the south-east of this area. Neither is drawn to a consistent

scale overall; the central portion of each is on a larger scale than the parts around the edges and covers a relatively small area. We have seen a similar technique on the Madaba mosaic, on the city-centred district maps from north Italy and on the late-medieval map of Dartmoor. The centre of interest of both maps lies in the western part of the Nanling mountains, the area between Canton and Kweilin, and both were probably drawn for military purposes: the area was on the frontier between a kingdom under the direct control of the Han emperor and a separate tributary state to the south. The map in ill. 58 pays particular attention to military features: the (approximate) rectangle outlining the central portion of the map seems to mark the limits of a defensive area with watchtowers on its borders, near its centre is a fortress that must have formed a headquarters, and lesser forts, storage depots and roads are also shown.

It is especially interesting that both seem to be a mixture of symbol-map and picture-map. On the map illustrated mountains are shown purely symbolically by the tendril-like lines, with clover-leaf and spiked protuberances, that dominate the whole map. On the other hand, the single triangles that mark the watchtowers and the triple triangles of the storage depots may be seen as a step towards pictorial representation, while the blue-green of the rivers and lakes and the outlined triangle with towers marking the headquarters provide a clearly pictorial element. We find the same pattern on the map of the larger region. There mountains appear as shaded outlines of snake-like appearance, while a sign looking like a set of organ-pipes may represent the nine peaks of a mountain or even the nine stone slabs in front of a temple. It may well be significant that on both maps (but rather less noticeably on the one in ill. 58) rivers are shown

58

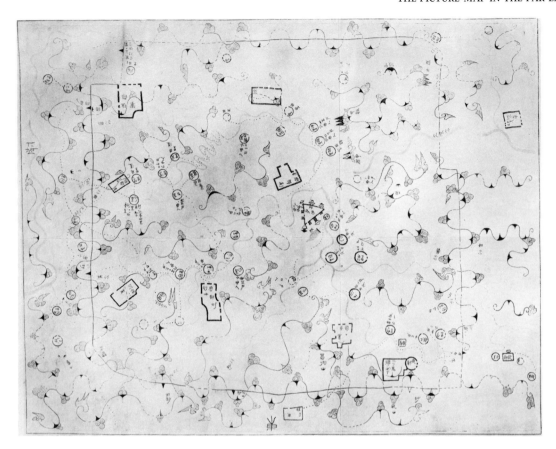

58 Reconstruction of a map of the 2nd century BC showing the mountains, rivers and military defences of an area in southern China, north-west of Canton. The original, drawn on silk and badly damaged along the folds, was one of three maps found in 1973 in a tomb near Chang-sha.

increasing in size as they proceed downstream, pointing to some sophistication of concept and technique; and significant also that features are represented differently on each – as the mountains are, or the roads which appear as simple lines on one map, dotted lines on the other – suggesting either an as yet undeveloped cartographic tradition or one developed so far as to allow room for diversity. The maps are some seventy years later than the earliest known reference to a map in Chinese sources: a map of the district of Tu-khang in 227 BC is mentioned by the historian of the next century, Ssŭ-ma Ch'ien.

It is interesting that these two surviving maps of the Early Han dynasty can be seen as military maps, for several of the earliest references to the use of maps in China come from accounts of military operations. The general who was waging a campaign against the Huns in 99 BC sent back to the emperor a copy of the map he had made of the areas beyond the imperial frontiers. And in 35 BC depictions of the storming of a Hun city in central Asia were given to the ladies of the court; there has been discussion whether these were maps or pictures, but as picture-maps they may well have been both. This use of maps for military purposes continued in later ages. The Emperor Hsien-tsung (805–20) used to consult a map set up in the imperial baths showing fortifications and strategic points north of the River Hwang-ho; this was by the geographer Li Chi-fu, whose son Li Tê-yü, as governor of Szechuan, made military maps for the work of controlling the barbarian tribes. It is an easy step from using military maps to using maps for general administrative purposes. Phei Hsiu – a key figure in Chinese cartography of whom we shall have more to say in chapter 9 – wrote in about AD 270 of local and general maps dating from the Later Han period (25–220) in official archives. It is told how in the late eleventh century ambassadors from Korea obtained maps from every local capital they passed through until the prefect of Yangchow, realizing the dangerous importance of what they were acquiring, took their collection and destroyed it.

From this and much similar evidence it seems that topographical maps were a normal instrument of government in China from at least the end of the third century BC, and that their use was not confined to the periods of unification under the Han and Chin dynasties but continued through the various divisions and fragmentations of the empire from the fourth century AD onwards. This is in contrast to Europe, where the use of topographical maps, at least for official purposes, failed to survive the decline of the western empire. That there really existed such a multiplicity of local topographical maps that they provided between them coverage of the whole of China can be demonstrated from the extraordinarily accurate maps of the whole country which we know to have been built up from them; we shall be looking at these in chapter 9 when we examine the introduction of surveying into local cartography. In two respects China from at latest the third century AD had advanced beyond medieval Europe in its mapping. In the first place there was a general awareness of the use and value of maps. And secondly the idea of representation to a consistent scale had been established; we can see the germ of this in the surviving Early Han maps in the way rivers are shown, wider near the mouth than near the source, but it may be that scale representation was achieved only in the mid- or late third century, for Phei Hsiu wrote disparagingly of the Later Han maps then surviving:

None of these employs a graduated scale and none of them is arranged on a rectangular grid. Moreover, none of them gives anything like a complete representation of the celebrated mountains and the great rivers; their arrangement is very rough and imperfect, and one cannot rely on them.

But for perhaps the next thousand years from Phei Hsiu's time topographical mapping in China stood at just about the same level as in sixteenth-century Europe, both in general awareness of maps and in mastery of the concept of scale (though not, be it noted, in surveying techniques).

Even if Chinese maps were now drawn to scale – and probably, as in sixteenth-century Europe, some were but many were not – this does not tell us what these maps actually looked like, just how they represented landscape. Here, as in medieval Europe, we suffer from a lack of published information; this may stem from a total loss of early topographical maps from local as well as national libraries and archives, or it may merely reflect scholars' preoccupation with the considerable achievements of the Chinese in the small-scale mapping of very large areas. But it seems at least very likely that the Chinese tradition of local mapping was predominantly pictorial. For one thing there is a pictorial element in some of the small-scale maps of large areas. Thus a map of the whole of China which was drawn in 1193 and survives in a copy carved in 1247 on a stone set up in the Confucian Temple at Suchow, shows mountains and forests pictorially, as in bird's-eye view. Dr Joseph Needham, incidentally, points out that the way it gives place-names in rectangular labels can be paralleled in certain seventh-century frescoes at Tunhuang in north-east China, which present a narrative in a landscape shown in bird's-eye view – a comparison that is interesting and suggestive in view of the associations we have seen elsewhere between picture-maps and picture-stories.

Certainly the Chinese local maps that we have from the thirteenth century on are entirely pictorial. Also from the Confucian Temple at Suchow is a plan of the city itself that was drawn in 1193 and carved on stone in 1229. We also have an early plan of Hangchow, originally published in a book of about 1274 but now known only from a re-issue of 1867 in which the map, printed from eight woodblocks, derives from an eighteenth-century manuscript copy. Both these city plans are clear picture-maps. They show in outline plan the streets and streams within the city walls, but other features are shown pictorially: walls, towers, gates, trees, the mountains outside the towns, and, on

59

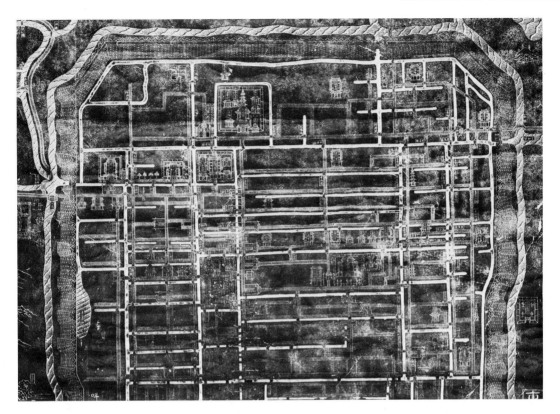

59 Part of a rubbing of a plan of Suchow, carved on stone in 1229. Pictorial details – important buildings, the city walls, the hills outside – are superimposed on an outline ground-plan.

the Hangchow map, ships in the harbour and the buildings of the imperial palace. The Hangchow map has pictures of individual houses only in the suburbs, but the Suchow map shows a few pagodas and courtyard houses in the town itself. From the sixteenth century onwards we have coastal charts, drawn to record defences and perhaps to aid navigation, that form a recognizable tradition: strip-maps, mostly looking from the sea in to the coast, with the few inland features – mountains, beacons, sometimes buildings – shown pictorially. The outlines of the coast and islands may ultimately derive from the older Chinese surveyed maps drawn to true scale, but by this time the concept of the scale-map had been lost, and there is great inconsistency of scale on the surviving coastal charts.

From the seventeenth century onwards we have a great profusion of Chinese picture-maps. A few examples will show their general character. Many were drawn for military or other official purposes. One military map of the late seventeenth century is a manuscript roll, some 11 feet long (3·5 metres), showing most of the length of the Great Wall with details of garrisons and of the barbarian tribes outside; scale and orientation both vary a good deal, but it gives notes of distances. Mountains and the Wall itself are shown pictorially, and other pictures include the flocks and tents of the barbarians. On this and other maps towns are marked by bird's-eye views of their walls and gates – a convention very like the European city ideogram that we saw in chapter 4. It appears on another eighteenth-century manuscript roll in a map of the thousand-mile course of the Grand Canal from Peking to Hangchow; here mountains, bridges and buildings all appear pictorially – the mountains, as often on Chinese maps, in green and bright blue with gold edging, the sea and rivers in green except the

VI

107

V

Hwang-ho (the Yellow River), which is appropriately painted yellow. Among other Chinese picture-maps are detailed maps of individual provinces, maps of military campaigns and maps drawn to form a permanent record of every journey of the emperor. Sometimes the detail shown is selective and conventionalized; sometimes the map is a realistic panoramic view. When in 1725 Fu Tse-hung produced his *Golden Mirror of Flowing Water*, a comprehensive account of the great rivers of China, he illustrated it with woodcut panoramas that are charming but quite unrealistic. In 1782 after severe flooding an agent was sent to the source of the Hwang-ho far away in the mountains of western China in order to propitiate the god of the river; the account of his mission was accompanied by a map of the area that is essentially a panoramic view of the mountains, showing in plan both the river itself and the plateau strewn with boulders where it takes it source.

In all it seems as if over at least the last five hundred years or so the native Chinese tradition of local mapping has been predominantly, even exclusively, a tradition of picture-maps. These seem to pay no more attention to measurement of distances or consistency of scale than the very similar sorts of picture-map produced in medieval Europe, and we are left wondering what became of the local measured maps which formed the basis for the small-scale maps of the whole country from the third century A D to the fourteenth, and which must have been produced not as a once-for-all operation but as a continuing tradition. It has been suggested that the replacement of the Yuan by the Ming dynasty in 1368 marked a decline in Chinese cartography and that there was a failure in tradition because maps were regarded as too highly secret to be copied freely. Some such explanation may lie behind the seeming retrogression from scale-maps to an entirely pictorial tradition of topographical mapping. But here again we may have more to learn from local maps (even relatively modern ones) surviving in archives and libraries in China: what is known to western scholars of local mapping in China comes largely from manuscripts and printed books preserved outside China itself.

Of one thing we can be certain: a continuing awareness in China of the use and value of topographical maps from the third century B C onwards, whether these were drawn to scale or were simply picture-maps. Moreover, this awareness came to extend to neighbouring countries too. We know of Chinese mapping in Tibet in the eighteenth century and in Outer Mongolia in the nineteenth. But Chinese cartography had spread abroad long before this. We have seen how in the eleventh century Korean ambassadors to China collected maps of the districts they visited; it might have been said that they were only following the example of an earlier Chinese ambassador to their own country who about 970 had brought back to China complete maps of Korea. And in Japan there is evidence of a continuous tradition of topographical mapping from the seventh century A D. It seems most likely that it was from China that the idea of using maps for official and other purposes came to Japan, but once established there topographical maps followed a quite independent course. There is no early Japanese equivalent to the use in China of local maps drawn to consistent scale. Equally there is no Chinese parallel to the very widespread use of printed maps among the Japanese populace at large from the seventeenth century to the nineteenth.

The first known reference to maps in Japan occurs in one of the two early-eighth-century books that are the oldest surviving works of Japanese history: the *Chronicle of Japan* (*Nihongi*), composed in A D 720. This mentions an edict of the emperor in 646 ordering that maps showing the boundaries of the provinces should be prepared and sent to the government. In 738 and 796 maps of provinces and districts, showing boundaries and the distances between posting stations, are said to have been made and in the eighth century too we have references to local maps, probably quite detailed ones showing the ownership of lands, that were used as a basis for taxation. None of these maps has survived, but some estate maps from the eighth century onwards have been preserved in temple archives and confirm the other evidence that topographical maps really were being made for administrative purposes at this period. The earliest known map of Japan probably also dates from the eighth century, though it is known only from much later reproductions; it was included in an encyclopaedia on court etiquette

60

compiled in the fourteenth and fifteenth centuries. It is attributed to a Buddhist priest,
Gyogi-Bosatsu, and it consists of a simple
outline showing the country's division into
provinces. Other somewhat later maps of the
whole country are more detailed. These maps
show Japan in recognizable shape, but they fall
short of the mastery of cartographic concepts
and the overall accuracy that we have reason to
attribute to Chinese maps of the same period.

From the seventeenth century on there is no
lack of topographical maps from Japan, including maps drawn to a consistent scale. The early
rulers of the Tokugawa Shogunate, which came
to power in 1603, ordered feudal lords to
produce maps of their areas, from which maps of
the whole country could be constructed. One of
these was drawn in the mid-seventeenth century by Hōjō Ujinaga on the scale of 1:432000
(about seven miles to the inch) from district
maps at twenty times this scale. In the University Library at Leyden is a collection of
twenty-one manuscript maps of the Japanese

60 One of the earliest surviving maps from Japan: a
plan of the 8th century AD showing a temple with the
estate belonging to it. Rivers, roads and the tree-lined
ridges of hills are marked.

provinces drawn at various dates in the seventeenth century but to a uniform scale, marking
the distances along roads and beside each town
and village giving the amount of rice it had to
pay in tax. A woodcut map published in five
volumes in 1690 of the Tokaido, the great tree-
lined highway 345 miles long between Kyoto
and Edo (Tokyo), was drawn to the scale of
1:12000 (about five inches to the mile) and was
based on a map drawn by Hōjō Ujinaga in 1651.
But it is drawn not in simple plan but as a
panorama – indeed it was designed by a notable
artist, Moronobu – and this is thoroughly
characteristic of Japanese maps of this period:
some were drawn to scale (many more were not)

61

109

61 Moronobu's panorama of the Tokaido, the road from Kyoto to Edo (Tokyo), was published in 1690. It provides an accurate picture of the road's course, enlivened by many figures of wayfarers and others, as in this waterside scene.

but the Japanese tradition of mapping from the seventeenth century on, as presumably in earlier centuries, was largely a pictorial tradition. Maps were the work of artists, and cartography was a branch of fine art. The development of topographical mapping was the development above all of the panoramic view.

We can see this in the maps of Japanese towns of the Shogunate period (1603–1868), though they were only one of the several types of printed map that were produced in quantity both for practical use as guides and for their decorative and artistic quality. Other types included maps of particular provinces, road maps and landscape panoramas. Probably the earliest printed city map was one of Hunai, the central part of Edo, that was published in 1632. This became out of date when a fire destroyed much of the city in 1657, and a new map, by Otikoti-Dôin, was produced on the government's orders and published in five sheets between 1670 and 1673; it was drawn to a consistent scale of 1:3250 (about twenty inches to the mile) and represented features by conventional signs instead of pictures. Although it was the basis for all subsequent maps of Edo down to the nineteenth century it was disliked by the public in general because it was not pictorial, and no similar map was made of any other city. However, plans of other cities were certainly produced showing the street pattern of the central areas in simple outline and restricting the pictorial elements to suburban areas; some indeed were drawn to a fixed scale, but this was evidently not thought very important and a stated scale was not always maintained over the entire map. And all these other town plans had a strong pictorial element in pictures of the temples, castles, bridges and so on. On maps of the larger cities, as Osaka and Kyoto, these pictures would normally be superimposed on what was basically an outline plan; maps of small towns were more likely to be simple bird's-eye views. We see the difference in the two woodcut plans in ills 62 and 63. One, published in 1689, is of Edo; it is basically an outline plan (though not drawn to scale) with some pictorial features superimposed. The other, published in 1813, is of Kōyasan, a small town that is a place of Buddhist pilgrimage; this is a realistic panorama showing every detail of the town and with a line of hills closing the picture at the top. In all, maps were published of some thirty-one Japanese towns in this period: the chief cities, spa towns, towns with shrines that attracted pilgrims. Few were produced of the castle towns, either for the sake of military security or simply because they would attract few visitors. The maps were often oriented so that the city's principal feature was at the top; at Osaka this was the castle that was the foundation of the town, at Kamakura, a centre of pilgrimage, the three chief shrines. The maps, which were produced for a wide market, were mostly coloured, at first by hand then, from the second half of the eighteenth century, by colour-printing; a plan of Nagasaki published in 1764 was the first Japanese map to be printed in colour.

But when we consider the Japanese maps of the Shogunate period we are faced with the problem of deciding how far we have a native cartographic tradition, how far one which at the very least has had grafted onto it elements imported from sixteenth- and seventeenth-century Europe. The Portuguese reached Japan in 1542, and although the Tokugawa Shogunate ended practically all contact with Europe until the mid-nineteenth century, there had been a period of some three generations during which Japan was exposed to the culture and technology of Europe, particularly of Portugal and the Netherlands. That an effective part of this exposure included map-making can be demonstrated by, for instance, the Japanese maps of eastern Asia that were drawn up very clearly on the model of Portuguese portolan charts. It seems likely that the Japanese tradition of topographical mapping was purely one of picture-maps, that the idea of drawing maps to a fixed scale was introduced to Japan from Europe and was not a native development. If we turn back to the town plans of the Shogunate we can see how several details of their history would accord with this, for some of the more advanced cartography of the seventeenth century gives way to less advanced concepts. We have seen how the most advanced of these city plans – that of Edo by Otikoti-Dôin – was produced early in the period, in the 1670s, and had no later parallel. The earliest plans of Kyoto were drawn to a consistent scale and the pictorial element was restrained; later plans had no fixed scale and had an abundance of pictures of temples, shrines, and landscape. Straightforward bird's-eye views, at first confined to smaller towns, came later to be made of the biggest cities as well – Kyoto in the late eighteenth or early nineteenth century, Edo in 1803. None of this constitutes proof, but it seems to fit better with the introduction of the scale-map from Europe during the period of early contact, than with the exploitation of techniques developed within the existing cartographic tradition. Without further evidence we cannot assume that Japan achieved independently the transition from the picture-map to the scale-map. In looking at the symbol-maps and the cartography of primitive peoples we saw how difficult it is to be sure that what seems to be a native tradition of map-making has not been created or at least affected by early contact with alien cultures; early topographical mapping in Japan shows how a similar problem can arise in the history of cartographical concepts and techniques in an advanced society.

62, 63 From the 17th century to the 19th there was a wide market in Japan for printed plans of towns, particularly of the largest cities and those that attracted pilgrims or other visitors. Shown here are two in contrasting styles. Above is the central portion of a plan of Edo (Tokyo), 1689: important buildings, bridges and some other features are shown by pictures, but they are placed on what is basically an outline ground-plan though not drawn to consistent scale. On the right is part of a plan of 1813 showing Kōyasan, a small town with a Buddhist shrine; it is a fully pictorial bird's-eye view.

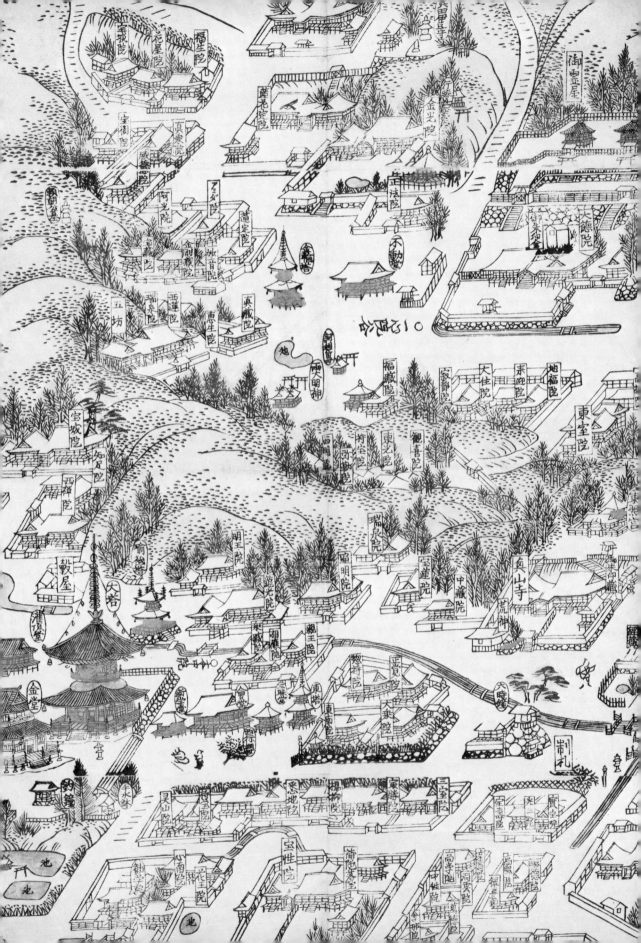

In China and Japan, then, we have traditions of topographical mapping which, while probably related in their origin, pursued quite different courses. We may reasonably wonder whether there were any similar traditions in the other cultures of the Far East; it seems likely, for instance, that topographical maps were made in south-east Asia, but no work on them has been published. In 1820 Francis Hamilton referred to what seems to have been a tradition of mapping in Burma:

During my residence in the Empire of Ava, from the 19th of March until the 27th of November 1795, I procured from the natives several maps of their country. These, as might be expected, were very deficient in accuracy; but I found the people wonderfully quick in comprehending the nature of our maps; and some of them, to whom I could render the occupation advantageous, very soon improved their plans, and produced drawings, which have tended to throw much light on the geography of . . . the *Farther Peninsula of India*.

Hamilton published several of the maps made by Burmese, but they were all of the improved variety learned under his tuition and they tell us little of the local tradition of map-making, though one, showing mountains by wavy lines topped with trees, is at least partly a picture-map. However, a map recently reported by Mr R. Kusmiadi of the Geological Survey of Indonesia shows what interesting material may still come to light. Dating from the fifteenth century and drawn in indigo on cotton cloth, it shows part of western Java and has the appearance of being, like the Early Han maps from China, on the borderline between symbol-map and picture-map. Some of its features are certainly pictorial – as the triangles that mark mountains – and the map is dominated by river-systems and by its many inscriptions. The map was discovered in a village in the Garut district of western Java; the villagers regarded it as sacred (along with a knife, a javelin and other objects) and a special ceremony was necessary before Mr Kusmiadi could handle or photograph it – a fact not without interest for the functions of maps among primitive peoples. From the Far East, as from practically every other part of the globe, we have more to learn of the early development of topographical mapping.

64   Part of a map of a district in western Java, drawn in indigo on cotton in the 15th century. The snake-like features are rivers; mountains are shown in triangular form. The inscriptions, in old Sundanese characters, identify the draughtsman as one Masjaya.

7

Mexico and India

ONE PART of Asia where we might well look for a tradition of picture-maps is of course India; and it may seem a little perverse to separate it from the other civilizations of the East and link it in a single chapter with Mexico. In fact the traditions of topographical mapping in both Mexico and India present the same problem. It is the problem we have just seen in looking at Japanese maps from the sixteenth century onwards but in a more acute form: the problem of deciding how far they are genuine local traditions of picture-maps, how far they have been influenced by ideas and maps brought in from Europe. From Japan we have not only clear references to the use of topographical maps but also actual specimens made long before the age of European contacts. But none of the picture-maps known to survive from Mexico or India is earlier than the sixteenth century, and many contain elements clearly introduced from Europe. From India at least earlier maps may still come to light; but at present we must be cautious in identifying and describing the local traditions of topographical mapping in both countries.

However, there seems no room for reasonable doubt that picture-maps were known in Aztec Mexico before the coming of the Spaniards. Hernando Cortés in the course of his conquest of Mexico wrote in 1520, in the second of his letters to the Emperor Charles V:

I also prayed Montezuma to tell me if on the sea-coast there was any river or bay where ships could enter safely, and he answered me that he did not know, but that he would have the coast drawn for me with its bays and rivers. . . . Another day they brought me a cloth, on which the whole coast was drawn, showing a river, larger than the others, flowing into the sea.

It has been remarked that the speed and ease with which this map was produced suggest that it was copied from one that Montezuma already had at hand. Again in 1526 Cortés wrote in his fifth letter to the emperor:

Some who had been in those parts described to me most of the villages on the coast as far as the residence of Pedrarias de Avila, Your Majesty's Governor in those parts, and they made me a drawing on cloth of the whole of it by which I calculated that I could go over the greater part of it.

And one of Cortés' companions, Bernal Díaz del Castillo, tells in his history of the Conquest how

In this pueblo of Istapa Cortés learned from the Caciques and native merchants all about the road we had to follow, and he even showed them a hennequen cloth which he brought from Coatzacoalcos, on which all the pueblos we should pass on the way were marked as far as Gueacalá.

We have substantial confirmation for the Mexicans' use of maps in several dozen surviving from the sixteenth and seventeenth centuries that are drawn wholly or partly in pre-Conquest style. Particular forms and conventions recur on them, suggesting that Mexico had a genuine tradition of picture-maps, like medieval Italy, and did not merely resort to them as an occasional expedient, like medieval England. However, we must remember that, as elsewhere, these picture-maps are the products of artists and their style and conventions belong to the culture's general artistic tradition rather than to a specifically cartographic one. Some of the Mexican conventions can be seen in ill. VII. This is a map of the valley of Tepetlaoztoc; it was drawn in 1583 to accompany a petition of the Indians of the area complaining of wrongs by Spanish colonists. Roads are shown by tracks of bare footprints between parallel lines, and ranges of hills in a style based on a simple sideways view: a wavy line along the top, a snout-like termination and along the bottom a straighter line with tiny rounded projections in groups of three. Both these conventions (the latter in varied forms) occur on other Mexican maps. Both are pictorial like other features on the Tepetlaoztoc map: the river (bottom right and along the top), the trees, the temple (upside down on the road that forms the map's left border) that marks the town of Tepetlaoztoc. The symbols – glyphs – in the panels like inverted shields represent place-names. Other maps give us further pictorial features: black dots for an area of sand, outline parallelograms marking the planks of a bridge, fish drawn in the rivers and so on. Whatever tradition of topographical mapping there may have been among the Aztecs it was indisputably one of picture-maps.

The Mexican maps in Aztec style mostly belong to one or other of two types. One consists of maps setting out the form and ownership of landed property or lordship. Boundaries are shown in red, by means of straight lines that most often meet at right angles. Sometimes the owner's claim to his land was supported by an elaborate genealogical tree, drawn on the map by means of lines linking tiny figures, each identified by a glyph. We see an example in ill. 65: a map drawn on cotton early in the sixteenth century to show the boundaries of the town of Metlacoyuca, north-east of Mexico City, and the successive lords of the area; it was found in a stone box at Metlacoyuca itself, in the ruins of the *teocalli*, or temple, that marks the town's site in the centre of the map. A number of the maps of this type were drawn for native lords quite soon after the Conquest to support claims to their hereditary estates after lands had been repartitioned. The other type is plans of towns. Some show little beyond a central open space and *teocalli*, but others show streets and watercourses, markets and squares. One, a large map, some 8 by 5 feet (2·4 by 1·6 metres), is believed to show part of the western suburbs of the city of Mexico. Apart from a few roads and waterways crossing the map diagonally it is made up entirely of some three hundred squares of individual plots of ground; on each is a human figure representing the owner and each is divided into thin strips, the so-called floating gardens, separated from one another by tiny waterways. Along one edge is a series of figures

showing the sequence of rulers; but an alien note is introduced by a picture of a church, Santa Maria la Redonda, built in 1524. On some less elaborate plans each division of a town is marked with a human figure representing its lord and giving his name. These town plans, as well as maps of wider areas, include some that Spanish governors sent back to the king in Spain; thus Philip II was sent plans of Misantla (on the Gulf of Mexico) in 1579, of Teozacualco (in the Oaxaca district of southern Mexico) in 1580 and of Epaçoyoca (in the region of Mexico City) in 1581.

Many of the maps contain features that depart from what seems to be the Aztec tradition and some have a mixture of Mexican and Spanish elements. Sometimes the imported features are obvious: the replacement of the *teocalli* by the Christian church with belfry, the translations given of names shown by glyphs. Sometimes they are more subtle; one likely innovation from Europe was the indication of direction, often shown by pictures of the sun in the east, the moon in the west. And the Aztec conventions for representing particular features were sometimes modified; a map showing the town of Tecamachalco and the genealogy of its rulers has all its figures sitting on mats instead of using this as a convention to distinguish the chiefs themselves, while a plan of Cozcatlan in 1580 uses tracks of horseshoes instead of human footprints to mark the roads. It has been suggested that the church school of Santiago Tlaltelolco in Mexico City was one channel by which Spanish art forms found their way into the tradition of Aztec mapping. That there was such a tradition of Aztec topographical maps there can be no serious doubt and because we have specimens made if not before at least very soon after the Spanish Conquest, its main features are fairly clear. But if we had only the maps drawn from the late sixteenth century onwards we would be hard put to it to distinguish the native Mexican from the imported European elements and might well

65 Mexican picture-map, early 16th century, typical of many in the lines of human figures representing the local lords' genealogy and in the tracks of bare footprints along the roads. The *teocalli* or temple at the centre marks the town of Metlacoyuca.

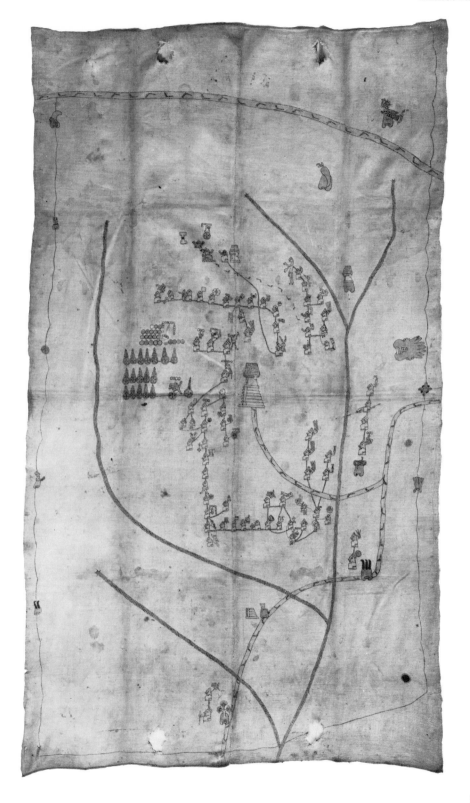

wonder whether there had been a pre-Conquest tradition of mapping at all. An instance of the sort of problem we meet on Mexican maps concerns the plans of buildings to be found on some of them: we see buildings in elevation round a courtyard, or a ground-plan with only doorways in elevation (like some Egyptian plans) or a ground-plan strictly adhered to throughout. One or more of these forms may belong to native Mexican tradition; alternatively they may all have come from Spain.

Like other picture-maps, the Aztec maps were only an aspect of their culture's general artistic tradition, and it is interesting that in Mexico this tradition contained two other types of representation that we have connected elsewhere with the emergence of picture-maps. One of these was the diagram with a topographical element like those from medieval Europe. In Mexico this took the form of a circle marking the principal town of a district, with around it other circles or segments marking dependent settlements; a human figure in the central circle, with a glyph giving his name, would represent the lord of the district. The other was the picture-story like those of the North American Indians. It is interesting that the only surviving pre-Conquest documents with any hint of cartographic content are of this sort. Two are early history books, both possibly of the fourteenth century: the area around the city of Tula appears in the Codex Zouche-Nuttall, the boundaries of the Toltecs in the Codex Vindobonensis. Another, the Codex Xolotl, is a copy of a record, drawn up before the Conquest, of the Chichimec tribes from their leader Xolotl down to the fifteenth century; the framework for its story is a set of formalized maps of particular areas and, interestingly, it uses series of footprints to show how the story moves, with the passage of time, from one scene to the next. And some of the post-Conquest maps contain references to particular incidents: the map of the town of Tecamachalco shows its lord standing, bafflingly, beside two temples in flames. There seems at least a strong possibility that the Aztec picture-maps took their origin in picture-stories, and behind these there may well have been an earlier tradition of symbol-maps as we saw among the North American Indians. To this development the diagram-map may have contributed a lot, a little

or nothing. Alternatively the origin of the Aztec picture-map may be entirely hidden from us.

In turning to topographical maps from India we should keep in mind the problems of Aztec mapping, for we have to answer the same questions that are posed by the maps from Mexico but in a more complicated form and from much less evidence. We have descriptions and reproductions of some very interesting Indian maps that seem to lie outside the tradition of European mapping, but they are few in number and late in date and are far from forming the substantial corpus of material needed to draw conclusions about the origins, development and style of Indian mapping. As in Europe we cannot tell whether early maps are waiting to be discovered and their significance recognized or whether they simply do not exist. At present we cannot say whether there were any real traditions of topographical mapping in India before the sixteenth century: the maps we have that contain elements of Indian style all date from periods when direct regular contact with Europe had been long established and we may merely be seeing Indian art forms applied to a type of representation introduced by Europeans. If, however, many more maps were to come to light so that we could establish that topographical maps were known in India before the period of European influence they would be of particular interest in view of the possibility of earlier influence from Chinese and Arab mapping.

Of possible Chinese influence we have just a hint in that several of the very few references to native mapping in India come from the northern frontier. Chinese interest and contacts in Tibet waxed and waned over the centuries, but at least provided a possible source of cartographic inspiration. It is interesting that the map in relief given to Warren Hastings – a sufficiently distinctive production to be specially described by Francis Wilford (see chapter 2) – was a map of Nepal. We are told how Francis Hamilton, when in Katmandu, in the early nineteenth century, was given five maps, drawn by Nepalese, of parts of Nepal and Sikkim; on his return he gave them to the library of the East India Company, but by 1876 they had disappeared. We have a description of a map of India made by a Nepalese showing mountains, rivers, roads, towns and temples. And from Tibet itself we have a strange, semi-cartographic picture of the great 66

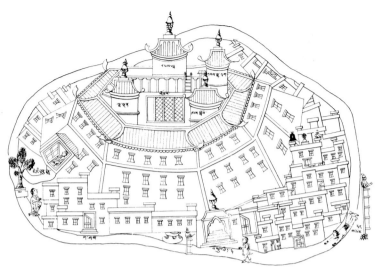

66 The great temple at Lhasa: a drawing of uncertain date. Tibet is a route by which knowledge of Chinese cartography may have reached India before the period of European settlement.

temple at Lhasa; it was published 'from a native drawing' in 1895, but otherwise we know nothing of its provenance or date.

The possible influence of Arab mapping is the harder to pinpoint in that we know so little of the form or development of topographical maps in Islamic countries. But we can find Indian maps that, like the early Arab regional maps, are scarcely more than symbol-maps with at most only vestigial pictorial features. Wilford wrote in 1805 that the Hindus

have also maps of *India*, and of particular districts, in which latitudes and longitudes are entirely out of question, and they never make use of a scale of equal parts. The sea shores, rivers, and ranges of mountains, are represented in general by straight lines.

Certainly we have maps that fit this description. One copied from an original of the late seventeenth or early eighteenth century covers roughly the area of modern Pakistan; roads and rivers are marked by single and double lines, places by various sorts of rectangles and circles, and only mountains are shown pictorially, as seen from the side. Again, moving to the Maratha area of south-west India, a late-eighteenth-century map of a much smaller area, covering some thirty-five miles of the coast to the south of Goa, consists of series of rough rectangles, showing rivers as double lines, forts as scallop-edged ovals; here the only pictorial feature is two fish drawn in the sea.

On the other hand early Indian mapping may well have included a tradition of picture-maps. Wilford's description of the Nepal map in relief shows that at least some of its detail was pictorial: rivers were blue, and trees were painted on the map. Early Indian literature contains a few references to what may have been picture-maps; thus the Abhijnāna Śākuntalam mentions what seem to be bird's-eye views. And from south-west India in the eighteenth century we have not only what are scarcely more than symbol-maps but also an interesting group of fully developed picture-maps which admittedly show European influence but which may also preserve features of a genuine Maratha tradition of mapping. Some of these maps cover very small areas; one is a plan of a garden, showing the trees and flowers to be grown there, and there are plans of forts; in ill. 67 we see one of Fort Vijayadurg (also known as Viziadrug or Gheria, on the coast between Bombay and Goa). We also have a map of a wider area: parts of the coast near Goa and of the rivers Kistna and Tungabhadra. This is oriented to the west, but the others in the group have east at the top and this may well reflect Maratha map-making practice; on the other hand the compass-rose that appears inside the fort on the Vijayadurg map is obviously European, as is perhaps the fact that the eight major compass-points are named around the border of this map. Other features that may come from a Maratha tradition of map-making include the placing of

119

67 Fort Vijayadurg on the west coast of India, 18th century. Some features, like the compass rose, must come from Europe, but the map as a whole may lie in a tradition of native Indian picture-maps.

wording so that the map has to be turned round to read it (as on ill. 67), single or double red circles drawn around place-names and contour-like lines to mark relief. But far more maps of this sort will need to be brought to light before the respective contributions of European and Indian cartography can be distinguished or, indeed, before we can speak with real confidence of topographical mapping in India before the age of European contact.

In India then it seems likely but not certain that there was a native tradition of picture-maps; in medieval Islam and Inca Peru the map modelled in relief may have been the nearest approach to the picture-map. With these exceptions we can say that we have found picture-maps in all the more advanced cultures of human history, though the evidence from medieval

Europe has shown us that we should always be wary of assuming a widespread tradition on the basis of a small number of surviving maps. The idea of making and using maps may have been restricted to particular areas or to limited classes of rulers, soldiers or officials more often than we might suppose. At the same time we can see clearly how very alike picture-maps are from all these societies, alike in concept and alike in technique. Representing landscape as though from an unattainable vertical or near-vertical viewpoint, drawn to no consistent scale but showing its features pictorially in the artistic style of the particular culture, seems to have been an all but universal stage in the development of cartographic concepts, one which sometimes evolved from the more primitive symbol-map, sometimes arose spontaneously. We must now turn to the third main phase of topographical mapping and see what societies have succeeded in taking the final step to true cartographic representation: the scale-map, which takes as its base a ground-plan accurately drawn to a consistent scale from a measured survey.

Part 3 : **Surveys**

8

The earliest scale-maps

IT IS an interesting comment on the unevenness of man's advance in ideas and techniques that when we turn to the most sophisticated form of topographical map – the scale-map, drawn to a consistent scale from actual measurements – we have to start by looking at some of the oldest maps that are known to exist. We have seen that the oldest picture-map known is a clay tablet of the third millennium from Nuzi, showing an estate or district probably in the same region of north-east Iraq. This tablet is unusual in its pictures of two ranges of mountains, but we have many other fragments of maps and plans on clay tablets from Mesopotamia, dating from the third millennium to about the eighth century BC. These mostly show simple outlines; the clay did not lend itself easily to the picture-map, and the pictorial element is restricted to the wavy lines that mark the water of a river or canal. Detailed work has been done on individual tablets, but no one yet has systematically studied these Mesopotamian maps and plans as a whole. This is a pity, for we might learn something of the way men first mastered the idea of drawing maps to scale. However, some light is shed on at least the chronology of their development by work that has been done on tablets bearing plans of one particular sort.

These are the plans of buildings. Twenty-one examples are known, and they have been listed and analysed by Dr Ernst Heinrich and Dr Ursula Seidl. Although they span fifteen hundred years or more – from the time of Naramsin, son of Sargon of Akkad, to the Late Babylonian period – they are extraordinarily alike in appearance: simple ground-plans, showing walls as double lines with piers and embrasures in outline, often with cuneiform characters giving the length of each wall and sometimes identifying individual rooms. Only one is markedly different: a Late Babylonian temple plan (in nine fragments and incomplete) on which the areas inside the walls are ruled out in squares to indicate floor-tiles. It is not always clear just what the buildings on the plans are, but some are temples, others dwellings built either in a solid block or else around a central courtyard. Where measurements are given we can tell whether or not the plan is drawn to scale, and here a clear division appears. In the oldest plans – those of the Akkadian period, in the third millennium – consistency of scale is disregarded, but probably all the later ones are at least meant to be drawn to scale. The earliest of these scale-plans is interesting on other grounds too, for it is not itself a clay tablet but rather a representation of one, carved in stone. It is held on the knees of a statue of Gudea, the ruler of Lagash in Sumer, some fifty miles north of Ur, whom we know from the tablets in his own archives to have been a great builder; he lived probably about 2100 BC, perhaps a century or so earlier. The statue is one of a group of statues of Gudea found at Lagash and now in the Louvre in Paris. There are no measurements on the plan, but it includes a scale-bar and although this is damaged its form can be reconstructed from the rule held by one of the other Gudea statues. The plan is peculiar in the complicated external shape of the walls and in lacking interior divisions; it may represent a temple or palace-hall, but we must remember that it is not itself a working plan but only a picture or impression of one and so need not be exact in its detail. No other of the plans has a scale-bar, but of the thirteen later than the Gudea statue all those (six in all) that bear measurements and are more than mere fragments seem to have been drawn to consistent scale. Indeed, the temple plan showing floor-

68, 6

68, 69 The building-plan above, carved on stone, lies in the lap of a statue of Gudea (left), ruler of Lagash in Sumer. Dating from the late 3rd millennium BC it is our earliest evidence for plans drawn true to scale: in the damaged corner of the tablet are traces of a scale-bar. Confirmation is provided by later actual scale-plans from Mesopotamia, drawn on clay.

tiles gives not just a conventional impression of the type of flooring, but is a scale drawing of the actual rows of tiles.

For want of similar analyses we cannot speak with the same precision about the other sorts of maps and plans found on Mesopotamian tablets, but they may well prove to bear out the evidence of the plans of buildings that the idea of drawing to scale was reached about the end of the third millennium, during the supremacy of Ur. Thus a tablet from Lagash dating from the late third millennium gives us a plan of fields with very full measurements but quite clearly makes no attempt at consistency of scale. On

124

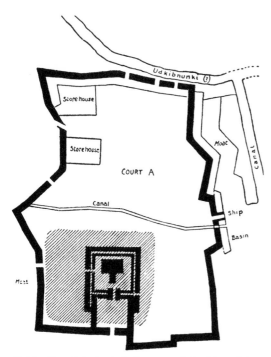

Store house

Store house

COURT A

Canal

Ship

Basin

Moat

Moat

Udkibnunki (?)

Canal

70, 71 That Babylonia had mastered the idea of the scale-map is shown by surviving plans, particularly of buildings, drawn on clay tablets. Here we can check the accuracy of the surviving fragment of a plan of Nippur, late 3rd millennium (left), against the outline revealed by excavation (above): it shows the site of the temple of Enlil in its sacred enclosure. In the centre is written Nippur's name: city of Bel.

the sale of properties, while by the twelfth century BC they are sometimes carved on *kudurru*, stones set up as a public statement of rights in particular lands. Of surviving fragments of town plans the Nippur plan reproduced is the most extensive, but others include two tiny portions of a plan of Babylon, one of them showing its southern citadel, the other the gate and suburbs west of the Euphrates. The third type of map known from ancient Mesopotamia – maps of large estates or districts, like the plan from Nuzi – is represented by fewer examples.

Given, again, the lack of any systematic study of these Mesopotamian maps and plans two further conclusions can only be regarded as very tentative. One is that the outline plans of every sort come mostly from the south, that is from Babylonia proper (Akkad and Sumer), Chaldaea and Elam, the regions south-east of modern Baghdad; far fewer come from the north, the original Assyria. This at least accords with the view of Babylon as cultivating the arts of peace, Assur those of war. The other is that the art of making scale-plans, or indeed any sort of outline ground-plans, vanished about the eighth century BC with the rise of the Assyrian Empire. It is picture-maps, not outline plans that might have been drawn to scale, that we find on the Assyrian bas-reliefs of the eighth to sixth centuries BC.

Certainly there is no trace of any link between the Babylonian plans and any later scale-maps and when the surveyed map next appears – in the Roman Empire – the idea had probably been reached independently of past precedent or outside influence. From ancient Greece, as we have seen, we have no topographical maps at all apart from the coin designs from Zancle and Ephesus. From Egypt we have enough picture-maps to suggest a real tradition of topographical mapping, but we have no surveyed maps. This is a little surprising, for we have working drawings in the form of elevations drawn true to scale on a background of ruled squares, and one might reasonably expect scale ground-plans as well. Moreover land in Egypt was being measured at a very early date – a treatise of the early second millennium deals with the problems of triangular and circular plots – and by the Ptolemaic period (fourth to first century BC) and

0, 71 the other hand a fragment of the plan of Nippur, some fifty miles south-east of Babylon, is drawn fairly accurately to scale, for it can be compared with the actual plan of the city as discovered by excavation; it dates probably from the late third millennium. At the same time the plan is not entirely accurate, and we cannot take it as certain that it really is a scale-map, based on actual measurement. It seems very likely that the ancient Mesopotamians applied the art of drawing to scale to topographical maps as well as to plans of buildings, but this has yet to be formally demonstrated. Certainly it is these two types of maps – plans of fields and of towns – that we would most expect to be drawn to scale. Plans of fields and small estates are fairly numerous from the third millennium onwards; the earliest are found on tablets that are deeds of probably long before, there was an elaborate system of taxation based on regular and very detailed surveys. But, as we have seen in medieval Europe, surveying is one thing and mapping another. There is nothing to show that the ancient Egyptians ever drew plans or maps of any sort to a fixed scale. Of scale-maps in the Roman Empire, on the other hand, we have very clear evidence indeed.

We have seen from rather scanty surviving examples how picture-maps were a familiar form of representation in the later Roman period, so much so that this was the sort of map used to illustrate the collected treatises on surveying. But from at least the first century AD there was a parallel tradition of surveyed maps. These were the maps drawn by the surveyors themselves, for, whatever the case in medieval Europe or ancient Egypt, in the Roman Empire mapping was regarded as part of surveyors' work. This is made clear from the treatises themselves. The surveyor had first to lay out lands for tillage and define the units of ownership. This might be done by the system known as centuriation: division into equal rectangular plots (*centuriae*: about 120 acres would be a typical size) aligned with mechanical regularity on two main axes (the *cardo maximus* and the *decumanus maximus*) which were used both as the bases of a system of parallel boundaries that might extend over many square miles and as points of reference to define the position of any particular plot. An alternative system was strigation, an older and less regular method of division. In either case once the lands and their internal bounds had been fixed the surveyors were expected to prepare a plan (*forma*). This would be sent to the imperial archives, and a copy also kept in the local administrative centre. Two of the surveyors' treatises – those by Frontinus and Hyginus, both written in the late first or early second century AD – say that the plans should be engraved on bronze, and the intention was that they should not simply be a transitory report on what had been done but a permanent record. It would not quite serve as a land-register for there was no machinery for keeping it up to date, but it would at least be a good starting-point for the settlement of any dispute that might arise.

Only one group of plans of this sort is known to us. They are not engraved on bronze, but carved on stone, and come from Orange (the Roman Arausio) in Provence. A few carved fragments had long been known, but between 1949 and 1955 a great many more were found, making it possible to reconstruct the overall form and pattern of the plans. The 443 fragments known seem to have come from three different plans. Each originally consisted of several rows of marble tablets which must have been fixed to a wall for permanent display. The plan best represented among the fragments (plan B) had at least four rows of these tablets and when complete it must have been over 18 feet high and 23 feet long (5·5 by 7 metres). Also found was an inscription apparently from the top of one of the plans stating that that particular survey was made in AD 77 on the orders of the Emperor Vespasian, so as to define and restore public lands that had passed into private ownership; it may have been part of the aftermath of the troubles of the year 69. Probably this came from the earliest of the three plans. It has been suggested that plan B is the second oldest, dating from the beginning of the second century and recording a repartition of lands over the same area, from Orange to Montélimar; and that the third plan, covering lands mostly south of Orange, is a little later still.

The plans carved on the Orange tablets were 72 of lands laid out by centuriation, rigorously maintained, and they thus consist of series of rectangles corresponding to the *centuriae* on the ground. Within each rectangle was carved its position on the grid in relation to the two main axes and a list of the plots of individual landholders within it, stating how each plot was held, what its area was and what payments were due for it. The centuriation proceeded undeviatingly across such obstacles as roads and rivers but these were marked on the plan in their routes through successive *centuriae*: roads and drainage ditches were shown by simple double lines, like the two main axes of the grid (the other bounds between *centuriae* were single lines), while rivers appear as double wavy lines. In places, particularly in covering the area near the mainstream of the Rhone, we are shown quite a complicated pattern of channels, meanders, confluences and islands. Given that the grids of regular rectangles on the plans correspond to grids of regular rectangles on the ground they can hardly help but be to scale. On

72 Fragment, carved on marble, of a plan of fields near Orange, laid out by centuriation, late 1st century AD. The double vertical lines down the centre mark one of the main axes of the field system; crossing it diagonally are two roads flanking a stream with an island in it.

the other hand on plan B the scale differs along the two axes: the squares of the *centuriae* appear as rectangles that are 14 centimetres high but only 12 centimetres wide, and thus the plan is on the scale of 1:5000 from east to west, but only six-sevenths of this from north to south. This may have been to limit the height of the plan so as to fit the wall where it was set up. But although the bounds of the *centuriae* were drawn to a fixed (if double) scale it need not follow that the exact widths and positions of the roads and watercourses within each *centuria* of the plan were determined by measurement on the ground. These seem rather to have been put in by eye, though of course their passage through the precisely plotted *centuriae* will have meant that their courses on the plan cannot have deviated far from their true position.

Confronted thus with a plan made up entirely of a grid pattern of rectangular pieces of land, drawn to a different scale along each axis and with the few other features simply sketched in, we might think that it has no very strong claim to be considered a scale-map; it is arguably little more than a topographical diagram. And given the Romans' predilection for rectangular planning – centuriated fields, grid-iron layout of towns and legionary camps – we might wonder whether the Roman surveyors' cartographical achievement was a very

significant one. But we are left in no doubt when we look at the only other substantial specimen of their work that survives to us, for this is one of the most impressive of all early achievements of topographical mapping. It is a large-scale plan not of a simple rectangular pattern of streets or boundaries but of the vast intricacies of the streets and buildings of Rome itself: this is the plan known to historians as the 'Forma urbis Romae'.

Like the Orange plans the 'Forma urbis Romae' was carved on rows of marble tablets and fixed to a wall. We even know which wall, for it still survives in the headquarters of the Franciscan Tertiaries at Rome; in the third century AD, when the plan was placed there, it formed part of the first-floor hall of the Forum Pacis. The plan was of enormous size, probably about 42 feet high and 60 feet wide (13 by 18 metres). It consisted in all of 151 tablets arranged in eleven rows; the four lowest were placed lengthways, the others alternately lengthways and upright. Of these tablets not one remains intact. What we have are 679 fragments, of which only forty-eight can be placed with certainty in their proper positions on the plan. They were not discovered in a single excavation but have come gradually to light in the course of centuries, mostly from around the base of the wall where the plan was fixed. In addition we have drawings made in the sixteenth century of a further thirty-three fragments that have since been lost. Altogether the surviving fragments and drawings cover no more than a tenth of the area of the whole plan and though we can reconstruct some of the parts that are missing (for instance by filling in the outlines of buildings that still exist) the parts of the plan of which we know both the appearance and the correct position amount to only about a twentieth of the whole.

From the buildings shown on the plan it can be dated between AD 203 and 211, and we know that the wall to which it was fixed was built about then. The plan seems to have covered the exact area of the city's limits at that time, except for two small pieces that it probably left out altogether. Almost certainly it served an official purpose, presumably connected with the business that was conducted in the hall of the Forum Pacis – but what that business was we do not know. The plan may somehow have been particularly connected with Lucius Fabius Cilo, who was Prefect of the city from about 205 to 208; he and his wife are the only two persons actually named on the known fragments, apparently to identify two properties that they owned. It is just possible that the plan replaced an earlier one in the same building, which we know was badly damaged by fire in 192. The plan's destruction or abandonment probably followed another fire which destroyed the hall where it was housed, possibly in 283.

It can be shown that the plan was carved after the tablets had been set in position. In some places there are relics of the original marking-out of the stone in the form of fine guidelines scratched on the surface; sometimes the carved line departs from these guides, and a few corrections to the carving can be found, but on the whole the plan seems to have been both marked out and carved with great care and accuracy. The appearance of the plan and the conventions it followed can be seen in the illustrations here. Walls are mostly shown by single lines, though on some important buildings the lines are drawn double and the intervening space hollowed out; this is not simply to show thick walls true to scale, for the hollowed-out line is sometimes impossibly wide. Single lines also mark certain boundaries.

73, 74

Dots mark columns or probably sometimes trees, and both dots within rectangles and small rectangles alone, either in outline or hollowed out, presumably mark columns standing on square bases. Larger rectangles and circles are used for the bases of statues, altars in front of temples, public fountains and so on, and a wide circle sometimes marks the precincts of a building. Some signs have not yet been satisfactorily explained; some may have stood for features in garden layouts which cannot be checked from surviving evidence on the ground. In only two cases does the sign used for a feature differ from its outline seen in plan: aqueducts (as in ill. 74) and other arches have the piers shown in plan as rectangles, but these are then joined with concave curves instead of straight lines, while sets of steps, sometimes shown in simple plan as cross-strokes between parallel lines, also appear as inverted V-shapes with or without cross-strokes (it is not clear whether there is any significance in these differences). But with these exceptions the entire work is in strict plan; for the most part it represents features by their outlines at two or three feet above ground level so that what we see is the pattern of ground-floor rooms, but some important public buildings are shown as though from above. All the principal buildings are named, and it is likely that the lettering was coloured – perhaps other parts of the plan as well. It is just possible that areas of the city that contained only private properties were left blank on the plan; this would explain why so many surviving fragments have no lines carved on them.

The orientation of the plan can be checked wherever an identifiable building or other feature occurs on a fragment with a straight side that shows it came from the edge of a tablet. In principle it seems to have had south-east (135 degrees) at the top: in practice the orientation varies between 130 and 136 degrees (which points to reasonable accuracy), except that at the temple of Claudius it reaches 144 degrees, suggesting that something went badly awry at that point. Its measurements point to a similar range of error. Dr G. Gatti has shown that its intended scale is 1:240, that is, in terms of Roman measurement, one *pes* on the map corresponds to the length of the *actus duplex* on the ground. Its actual scale is very slightly

73, 74 Fragments of the scale-plan of Rome, AD 203–11, that is the most impressive monument we have of the work of Roman surveyors. Carved on marble, to the scale of 1:240, it was set up on the wall of a public building. Showing nearly all features in strict outline ground-plan, it is remarkable both in the sophistication of its cartography and in its general accuracy.

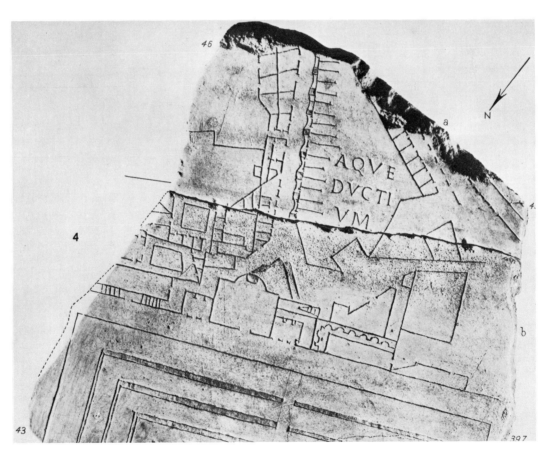

smaller. As the plan is so fragmented it is not easy to check lengths, but Dr Gatti was able to establish the lengths on the plan of thirteen substantial distances; if the scale is taken as 1:245, eight of these measurements are within 3 per cent of being correct, while the other five show variations of from 5 to 17 per cent. As what we have is not the surveyors' finished drawing but a copy that must have suffered some loss of accuracy in being set out and carved on a large vertical surface, these calculations point to fairly exact surveying. One interesting feature of Dr Gatti's figures is that the longer lengths are more accurate than the shorter ones: of the eight that are over 1200 yards (1100 metres) only one is more than 3 per cent out. This may be because the surveyors' inaccuracies one way or the other in their measurements simply cancel out over the longer distances. On the other hand it may point to an accurately constructed framework for the whole plan, with a less exact filling in of detail within its individual sectors; but how this framework was made we do not know. We know of some Roman surveying instruments, but these include none for measuring angles. If in fact the plan was as accurate overall as Dr Gatti suggests we do not know how this was achieved.

But the plan of Rome shows clearly that, however they did it, Roman surveyors were capable of constructing maps of great complexity to a consistent scale. Both the concept and the techniques of the scale-map had been thoroughly mastered, so that in the Roman Empire there were two parallel traditions of map-making, one of picture-maps and one of scale-maps. The union of surveyor and map-maker, which was to elude medieval Europe (except perhaps in the cities of Italy) had been achieved, just as it had been achieved some two thousand years before in Mesopotamia. But compared with the plan of Rome and even the plans of centuriation from Orange, the only other relics we have of this achievement are unimpressive. At the same time they confirm that the scale-map was in regular use in the late Roman period. All are entirely in plan, with no pictorial features. One is a mosaic, of which three fragments survive, giving a plan of a set of baths on the scale of 1:16 (that is, one *digitus* on the map represented one *pes* on the ground). The other five were all carved on marble (of one,

75, 76 The only surviving fragments of two carved scale-plans by Roman surveyors. They clearly belong to the same tradition as the 3rd-century plan of Rome (ills 73, 74) while differing in some details, notably in showing walls by double lines instead of single. The fragment above, found at Rome, shows private buildings inscribed with their owners' names. The one below, from Ostia, seems to show workshops; the carved letters may be measurements or the numbers of the city blocks.

75, 76

now lost, our only record is a seventeenth-century drawing) and only two are complete – of the rest we know only fragments. They all show buildings: private houses, workshops and perhaps tombs and gardens. In general appearance they are very like the plan of Rome. We find, for instance, steps shown in the same way with cross-strokes between lines that are parallel or converge and on one of the plans there are regularly placed dots that can only be taken as trees. On three of them, however, all walls are shown as double lines instead of the single lines that predominate in the plan of Rome, and this must have been an alternative convention. Several of them also differ from the Rome plan in having measurements entered on them. One of them certainly and possibly two others as well are on the same scale as the plan of Rome, 1:240. Two of the others use different scales from one part of the plan to another so that although they clearly belong to the same tradition of map-making their claim to be considered real scale-maps could be questioned; it may be that they were not actual plans but rather pictures of plans, carved on monuments, like the plan held by the statue of Gudea.

We saw in chapter 4 how Charlemagne in his will bequeathed panels engraved with representations of Rome and Constantinople, and that these may have belonged to either tradition of Roman mapping, deriving from either the picture-maps or the scale-maps of the late classical period. We have in fact two surviving witnesses to continuing knowledge of at least the form, if not the techniques, of the Roman surveyors' maps after the period of the barbarian invasions. One of these is the plans that accompany the account of Arculf's journey to Jerusalem. Arculf was a Gallic bishop who travelled to the Holy Land in 670; on his way home his ship was carried away to Iona, off the island of Mull on the west coast of Scotland, and here he was cared for by Abbot Adamnan. It was for Adamnan that he wrote the account of his journey, and the text that survives in various copies from the eighth century onwards is in effect Adamnan's edition of Arculf's account. In some manuscripts of the ninth century and later the text is accompanied not only by the diagrammatic plan of Jerusalem already mentioned in chapter 4 but also by plans of four buildings: the Church of the Holy Sepulchre,

the Church of Sion, the Church of the Ascension and a structure at Shechem ('Sichem Jacobi': Neapolis, now Nablus). All are simple outline plans with no pictorial elements. Walls are shown by double lines. We cannot tell whether these plans were really the work of Arculf, but whoever drew them was working in the stylistic tradition of the Roman surveyors.

So too was the draughtsman of a very much better-known work, the monastic plan from St Gall in Switzerland, which has long been a subject of research and debate among scholars. This dates from the early ninth century: a dedicatory note says that it was being sent to Gozbert, who was Abbot of St Gall from 816 to 837. Unfortunately it does not name the sender, but as he addresses the Abbot as 'sweetest son' ('dulcissime fili') he must have been someone highly placed in church or state and it seems most likely that he was Heito, who from 806 to 823 was both Bishop of Basle and Abbot of Reichenau; from the handwriting of the two scribes who wrote its inscriptions it has been shown that it was at Reichenau that the plan was drawn. What the plan shows is the layout of a monastery. This may have been intended as an actual building-plan, perhaps to be connected with the synods for monastic reform that were held at Aachen in 816 and 817. Alternatively it has been seen as an idealized scheme for setting out a monastery, almost a planner's dream of how to fit together its various components. The plan has a great many inscriptions explaining what the various buildings and rooms were to be used for, but its basis is a simple outline which takes us straight back to the 'Forma urbis Romae': walls are shown by single lines, the piers of the church nave by dots within squares, and so on. At the same time it departs from the strict outline ground-plans of the Roman surveyors and a few features appear in elevation. Thus around the main cloister and elsewhere arches and arcading are shown by semi-circles, singly or in series. The two round towers flanking the west end of the church are drawn not as ring outlines but as spirals, which may be meant to represent spiral staircases or may be simple embellishment. Fruit-trees in the cemetery are marked not by regularly spaced dots but by formalized pictures of saplings with tendril-like branches. Whether on his own initiative or following existing precedent the

77

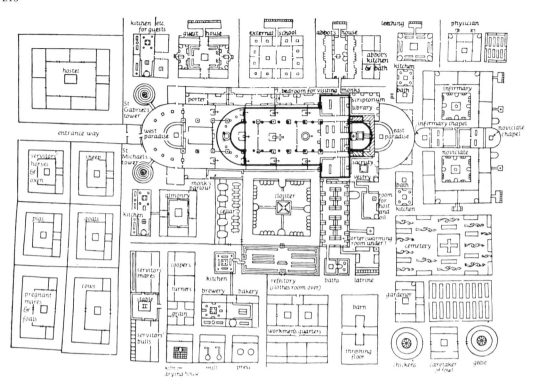

draughtsman has modified the concept and tradition of the Roman surveyors' scale-maps.

It seems likely that the St Gall plan was envisaged as a scale drawing: it has been demonstrated by Dr Horn and Dr Born that it was drafted on the basis of a square grid and that the dimensions of the church as given in the inscriptions can be seen as consistent with a value of $2\frac{1}{2}$ Carolingian feet for each square of the grid, giving an overall scale of 1 : 192. It may be that Arculf's plans also were drawn to scale, but we can hardly determine this for the surviving copies must be at several removes from the original drawings. Certainly besides their common ancestry in the Roman surveyors' plans there is a detail that links the St Gall plan and Arculf's drawings: doorways are marked not just by gaps in the walls but by short transverse lines (double on Arculf's plans, single on the St Gall plan, like the walls themselves). These may have originated in tiny marks representing jambs or pediments, but

77 The last map known in the tradition of the Roman surveyors: a simplified drawing, with Latin names translated, of the plan of a monastery drawn in 816–37 for the abbot of St Gall. Its author and its purpose – whether actual building plan or theoretical schema – are uncertain, but it seems to have been drawn at Reichenau and was intended as a scale-plan.

they seem to have become a conventional sign. There is nothing similar on the 'Forma urbis Romae' or other Roman plans: an occasional thickening of line beside a doorway is probably merely an accident of carving. This may seem a small point, but it is an important one: it links our two post-Roman works and suggests a continuing tradition that originated in the Roman surveyors' plans but developed some features of its own. But the St Gall plan must have been one of its last products. We find no later traces of the scale-maps of the Romans.

9

From itinerary to survey

IN THE last chapter, under the heading of the earliest scale-maps, we looked at the maps made by the surveyors of ancient Mesopotamia and imperial Rome. This accords with what evidence we have, but it should be said that the Roman surveyors win their place there only by a very short lead. In China maps drawn to consistent scale from measured survey were being constructed certainly by the third century AD and just possibly somewhat earlier. And whereas we do not know how the third-century Roman surveyors constructed their plan of the city of Rome, we have a remarkably clear account of the way scale-maps were being made in contemporary China.

This was written by Phei Hsiu, already mentioned in chapter 6, the most distinguished of China's early cartographers. Given responsibility for imperial works in AD 267 by the first of the Chin emperors he investigated existing maps of the various parts of China and then, finding them inadequate, constructed a new map of the whole empire in eighteen sheets. His map is lost, but we have his preface to it in which he tells how it was made. This is of such extraordinary interest that it must be read in full:

In making a map there are six principles observable:

(1) The graduated divisions, which are the means of determining the scale to which the map is to be drawn.

(2) The rectangular grid, which is the way of depicting the correct relations between the various parts of the map.

(3) Pacing out the sides of right-angled triangles (*tao li*), which is the way of fixing the lengths of derived distances (i.e. the third side of the triangle which cannot be walked over).

(4) Measuring the high and the low.

(5) Measuring right angles and acute angles.

(6) Measuring curves and straight lines. These three principles are used according to the nature of the terrain, and are the means by which one reduces what are really plains and hills (literally cliffs) to distances on a plane surface.

If one draws a map without having graduated divisions, there is no means of distinguishing between what is near and what is far. If one has graduated divisions, but no rectangular grid or network of lines, then while one may attain accuracy in one corner of the map, one will certainly lose it elsewhere (i.e. in the middle, far from guiding marks). If one has a rectangular grid, but has not worked upon the *tao li* principle, then when it is a case of places in difficult country, among mountains, lakes or seas (which cannot be traversed directly by the surveyor), one cannot ascertain how they are related to one another. If one has adopted the *tao li* principle, but has not taken account of the high and the low, the right angles and acute angles, and the curves and straight lines, then the figures for distances indicated on the paths and roads will be far from the truth, and one will lose the accuracy of the rectangular grid.

But if we examine a map which has been prepared by the combination of all these principles, we find that a true scale representation of the distances is fixed by the graduated divisions. So also the reality of the relative positions is attained by the use of paced sides of right-angled triangles; and the true scale of degrees and figures is reproduced by the determinations of high and low, angular dimensions, and curved or straight lines. Thus even if there are great obstacles in the shape of high mountains or vast lakes, huge distances or strange places, necessitating climbs and descents, retracing of steps or detours – everything can be taken into account and determined. When the principle of the rectangular grid is properly applied, then the straight and the curved, the near and the far, can conceal nothing of their form from us.

Different opinions have been put forward on the exact meaning of some of the terms that Phei Hsiu uses, but the general picture is clear. The basis of the map was a grid of equidistant lines forming a network of squares. Dr Needham discusses, interestingly, how this idea may have arisen; possible sources include the boards used for divination and the warp and weft of the silk on which (as we have seen) some early Chinese maps were painted. Features were placed in their proper position on the map simply by measurement. If natural obstacles made it impossible to measure a particular length it would be calculated by making it a side of a right-angled triangle and measuring the two other sides. The reference to acute angles in the fifth principle must mean the measurement of angles so as to enter correctly the direction of roads and rivers in relation to the grid. But there is no reason to suppose that measurement of angles played any part in constructing the map as a whole. One way of making a scale-map is by triangulation, using angle-measurement to fix accurately the positions of a few key points to form a framework into which the detail can be fitted. This is to start from the outside, as it were, and work inwards. But this was not Phei Hsiu's method. Rather, he could be said to have worked from the inside outwards, the map of the large area being constructed from a patchwork of local maps. Harking back to Ptolemy's distinction between the geographic and the topographical map we might say that, even though it covered such a great area, Phei Hsiu's map of China was not really a geographic map at all, but a vast compilation of topographical maps (though it is fair to add, in passing, that the maps attributed to Ptolemy himself must have been made by methods that were not fundamentally different). Phei Hsiu does not describe the grid he used, but it presumably consisted, as in later Chinese maps, of horizontal parallels running due east and west crossed by others at right angles. How far north or south places lay may well have been checked by simple astronomical observation, such as measuring the noon-tide shadow of a stick of fixed length; but dead reckoning,

simple measurement of distance, would be the only way of fixing positions in relation to the vertical lines of the grid.

If we do not have Phei Hsiu's own map we have at least later maps based on rectangular grids that must have been made by the same methods. Two are outstanding. One is a map of the whole of China which probably dates from the eleventh century, but which we know only from a copy carved on stone in 1137. An inscription tells us that each division of the grid is 100 *li* (a measure of about one-third of a mile or half a kilometre). The map is not faultless, but its general accuracy, particularly in the river system, places it among the world's great cartographic achievements. The other is an atlas of China and its neighbouring lands that was published about 1555, taking as its basis a large map of China, now lost, that was drawn by Chu Ssu-pên in the second decade of the fourteenth century. The preface to the atlas explained how it was constructed on a grid: 78

Chu Ssu-pên's map was prepared by the method of indicating the distances by a network of squares, and thus the actual geographic picture was faithful. Hence, even if one divided the map and put it together again, the individual parts in the east and west fitted faultlessly together. . . . His map was 7 feet long, and therefore inconvenient to unroll; I have therefore now arranged it in book form on the basis of its network of squares.

The atlas contained a general map of China at a scale of 400 *li* to each division of the grid, and maps of individual provinces and regions, including special maps of the River Hwang-ho and of the Grand Canal; most were drawn with grids of 100 *li* to the division, but some used grids of different scales, from 40 to 500 *li*. 79

The carved map of 1137 and the atlas of 1555 both show that Phei Hsiu's method was a perfectly workable one, capable of producing extremely good maps of wide areas. They are a tribute both to the quality of the surveying that lay behind them, and to the advanced ideas of mapping in China, whether maps of this sort were widely used or – as seems likely – confined to the official and military classes. And the fact that it was possible to produce these maps at all shows quite clearly that there must have been, everywhere in China, topographical maps of individual districts, drawn accurately to scale

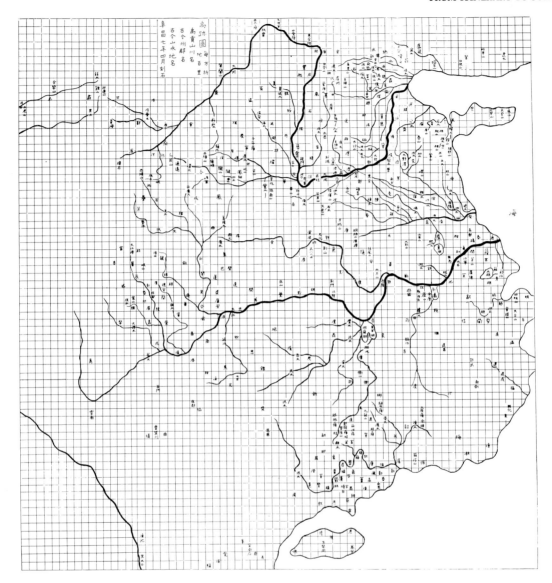

78   This map of China was carved on stone in 1137. It is our most striking monument of the tradition of Chinese maps drawn to scale that had emerged by the 2nd century AD and was probably lost by the 14th century.

from measured surveys. Although these maps have apparently all disappeared, leaving no other trace, they were presumably in normal use at least from the third century to the fourteenth, from the time of Phei Hsiu to the time of Chu Ssu-pên. The scale-map then seems to have declined and vanished, leaving only the tradition of picture-maps that we have already discussed.

Essentially Phei Hsiu's method was that of the measured itinerary. Route after route – by road, by river, by canal – would be followed, its

length and changes of direction duly measured and noted. And we can see that if this was done with complete accuracy, for crossroads as well as for those radiating from a single place, a perfect scale-map could be constructed. In practice it would involve some trial and error,

135

some juggling with the lines of routes to make them fit exactly on the map. But the result would be a map built up from lengths and angles measured solely along the lines of routes. Turning to medieval Europe it is interesting to find map-makers gradually feeling their way towards the scale-map, the map based on measured survey, by just this same way of the itinerary map, the map of routes.

A map of a single route is just about the simplest map there is. As we saw from Robert Cole's survey of Gloucester in 1455, a list of places (in that case houses) arranged in the order they are met with on the ground is all but an itinerary map in itself. The Roman Empire was certainly familiar both with written itineraries and with itinerary maps. Vegetius, in his treatise on warfare written in the late fourth century, says of the general in charge of a military expedition:

79 Though published in 1555, the atlas from which this map is taken was based on the maps drawn by Chu Ssu-pên in the early 14th century. We see here the general map of China, bounded on the north by the Great Wall and the Gobi Desert; its other maps are of the individual provinces of China and neighbouring lands. Like the carved map of 1137 (ill. 78) it is based on a measured grid, the system laid down by Phei Hsiu in the 3rd century AD.

First of all he should have itineraries, written out very fully, for all the areas where the war is being fought, so that he may know not only the distances between places but also the state of the roads, and so that he can keep in mind the short cuts, the side roads, the mountains and the rivers, all faithfully described. And indeed it is said that the most skilled commanders have had itineraries not just written but drawn ['non tantum adnotata sed etiam picta'] for the provinces where their work lay.

It was certainly not only for military use that the Roman Empire made itineraries. Those that are known to us were made primarily for civil administration. It has been suggested that they all have a common origin and that this was a now lost route map of the whole empire drawn up in the reign of Caracalla (211–17). We have three very full written itineraries: the Antonine Itinerary (late third or early fourth century), the Bordeaux Itinerary (of a journey said to have

137

been made in 333) and the Ravenna Cosmography (compiled about 670). These are simply lists of places along routes, giving the distances between them. In the Antonine Itinerary all distances on land are in Roman miles, those by sea in *stadia*; the Bordeaux Itinerary (from Bordeaux to Jerusalem and back) gives distances from Bordeaux to Toulouse in leagues, all others in miles, and when it reaches Palestine it expands into a brief guide-book. The sixth-century Madaba mosaic was probably based on a set of itineraries or even an itinerary map, for all the places that are named on it, apart from those taken from the Bible, lie on main roads. We see a faint hint of an itinerary map in the Rudge Cup, a small enamelled bronze bowl made in Britain in the mid-second century. Around the rim it has the names of five places along Hadrian's Wall, almost certainly taken from an itinerary (four of them occur in reverse order in the Ravenna Cosmography); below is probably a representation of the Wall itself, a design of battlemented towers with straight walling between. It was probably one of a mass-produced set on which all the forts of the Wall would be named, being made with an eye to the market provided by legionaries stationed there.

Besides written itineraries we have two quite clear itinerary maps from the Roman Empire. One is a fragment, about 7 by 18 inches (18 by 45 centimetres), of a piece of parchment that originally covered a shield. It was found in excavations on the site of Dura-Europos on the upper Euphrates in Syria, and dates probably from the first half of the third century. A thin white line divides it into two sections. The left side is painted blue to represent the sea – the Black Sea in fact – and on it are shown one rigged ship and part of another, and perhaps two small boats as well. The right side, painted with a red background, represents the land, and along the line of the coast a dozen places are named in Greek, each followed by a note of its distance from the next; beside each is a picture of a small building, and blue lines mark rivers that cross the route. All but the last two of the places named lay on the west coast of the Black Sea, starting at Varna in Bulgaria, so that we might seem to have a fairly straightforward picture-map with south at the top; but the two names at the foot of the fragment are Kherson (at the mouth of the Dnieper) and, probably,

Trebizond (on the north coast of Asia Minor). We can only guess what the completed map looked like, but it probably showed a single itinerary, perhaps a route followed by the shield's owner, that began and ended in Syria. Although the names are in Greek they are almost certainly translated from a Latin itinerary: the River Danube is marked and named twice, once with its Greek name, once in a Greek transliteration of its name in Latin.

Our other Roman itinerary map is on an altogether different scale of complexity, for it covers the whole empire. We know it only through a copy (but apparently a faithful one) that dates from the eleventh or twelfth century; this copy belonged in the sixteenth century to the great Augsburg antiquary and collector Konrad Peutinger, after whom it is known as the Peutinger Table. In the form that was copied the map apparently dates from the fourth or fifth century, but this was probably a revised version of a rather older original. It consists of eleven sheets of parchment – originally there were twelve, but one is missing – which, put together, form a strip some 23 feet long but only 14 inches wide (6·75 by 0·34 metres). From the small portion reproduced we see at once that in form and concept it is very different from an ordinary map. It has in fact a single aim: that of showing the places (and outstanding natural features) along every road, branch or crossroad in the order the traveller will come to them. This aim it performs faithfully, but that is the beginning and end of its cartographical function: the representation of relative distances and directions lies beyond its scope. We see this clearly in the illustration. At the top is the River Rhine, flowing through (on the right) Lake Constance with a row of trees representing the Black Forest above. Below it the line of the Alps sweeps across the map, meeting on the left another mountain chain, the Ligurian Alps and the Apennines; between them lie the west end of the River Po and its tributaries, and places marked include Aosta (Augusta Pretoria), Turin (Augusta Taurinorum) and Genoa. So far the general shape of the map has not diverged far from what we might expect; but setting out from the Ligurian coast we cross no more than a narrow channel dotted with islands (the Balearics, Corsica and Sardinia) before we reach land again, a long thin arm that is the Iberian

80

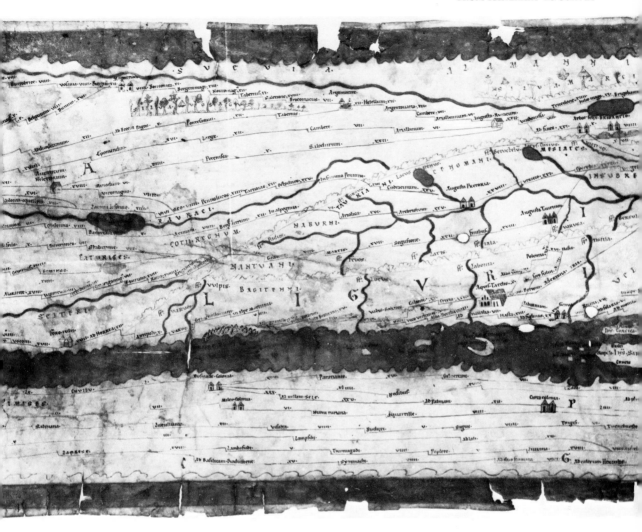

Peninsula. It is reached on the map by following the roads left through southern Gaul, then back across the Pyrenees into Spain; every route passes through the right places in the right order, and other spatial relationships are irrelevant to the map's purpose. In this way it is possible to cover the whole empire in this long thin strip. Sections of the map to the right of this one show not two but three narrow strips of land: Spain at the bottom, then Italy and at the top the Balkans. Within its terms of reference the map is very detailed. It shows a great many roads and it gives the distances along them, mostly in Roman miles, but changing to local measures in Gaul (leagues) and Persia (parasangs). Towns or other settlements or staging-

80 Part of the Peutinger Table, the late-Roman itinerary map that survives only in this medieval copy. While its outlines bear little relation to the map of Europe we know, it is an accurate, detailed and efficient guide to routes and distances.

posts are mostly marked in three ways (which all appear in the illustration): twin towers, a single large building or (and these are mostly places named *aquae*: spa) a more elaborate complex. The exact meanings of these conventional signs are a matter of debate. But though distances are marked there is no attempt to show them to any fixed scale on the map – indeed the distortions in its overall shape would make this scarcely possible.

The itinerary maps, then, while recording measured distances, seem quite unconnected with the tradition of surveyors' maps in the Roman Empire. The ships and houses on the shield-map, the trees marking forests, the lines of mountains as seen from the side and the buildings marking towns on the Peutinger Table all link them rather with the Roman tradition of picture-maps. It is very interesting that itineraries and itinerary maps should have been, so it seems, in common use in the Roman Empire, but it is no less interesting that so elaborate, and indeed so sophisticated a product as the Peutinger Table should have been carefully copied in the middle ages. For from medieval Europe too we find both written and drawn itineraries. A particularly well known one is the Bruges Itinerary of the fourteenth century: lists of places along the roads from Bruges throughout Europe (except Britain and Portugal), giving distances by local measures (leagues in France, miles in Germany, days' journeys in Hungary and Greece, and so on). It is perhaps surprising that it omits Britain, for, whether from chance survival or local tradition, it is from England that we have most evidence from the middle ages before the fifteenth century for the use of itineraries.

There are English written itineraries. An early example is the record of the route of Sigeric, Archbishop of Canterbury, from Rome to the English Channel after he had gone to receive the pallium from the Pope in 990. Another is a list of between 1405 and 1408, written at Titchfield Abbey in Hampshire, giving the routes and distances between Titchfield and the other Premonstratensian monasteries in England. But it is the itinerary maps from medieval England that are most distinctive. Several of the local maps discussed in chapter 5 are really itinerary maps, for they are concerned with spatial relationships only along a single line, as, for instance, the early-thirteenth-century plan of the springs and pipes at Wormley. The most notable example is a map of Sherwood Forest, dating from the late fourteenth or early fifteenth century. It consists of little beyond lists of names of villages and other features lying on separate routes through the Forest. The order of the names and the general direction of each route is quite accurate, but the map would be a most misleading guide to

direction and distance between places lying on different routes: it is a true itinerary map. An English itinerary map for a much longer journey is the itinerary from London to Apulia that Matthew Paris, the St Albans monk, historian and artist, drew up (in four successive versions) in the mid-thirteenth century.

But this was not Matthew Paris's only itinerary map. There is a strong element of the itinerary in the more advanced and workmanlike of his two maps of the Holy Land – the one of which we have only one copy, probably a draft (see chapter 3) – in that between the names of towns along the coast are notes of their distances apart: 'one day's journey', '6 leagues' and so on. Moreover, Matthew Paris's map of Great Britain can also be seen as an itinerary map. It exists in four versions, and it is the one reproduced here that shows most clearly how it is based on an itinerary. We can see at once that the backbone of the whole map is a route from Dover to Newcastle upon Tyne by way of London, St Albans, Belvoir (where there was a cell of St Albans Abbey), Pontefract, Durham and ten other named places. This route is shown as a straight vertical line, and towns, counties, rivers and coastlines elsewhere in the country are placed in their relationship to it rather than to each other. This does not explain all the map's eccentricities – Colchester, for instance, is placed to the west of the basic route instead of to the east – but it does help us to see the idea behind the map. Professor Richard Vaughan has remarked that showing this itinerary as a straight line is 'the one major error in the maps'; but this is perhaps applying the wrong standard. It may not have been a mistake at all. Matthew Paris may well have shown the route from Newcastle to Dover as a straight line while knowing perfectly well that it makes a right-angled turn at London to run east instead of south. Because he wrote 'Auster' (south) at the foot of his map it need not follow that this was to show how all its features were aligned – it may

81 Map of Britain by Matthew Paris, mid-13th century. On England's northern border are Hadrian's Wall and (misplaced) the Antonine Wall; below them, in vertical column, is the line of places that form the map's backbone, from Newcastle ('Nouum castrum') to Dover.

81

48

have shown no more than the general drift of the map's layout (and, after all, if one wanted to go from London to lands south of Britain it was by way of Dover that one would go). It would not have been unreasonable for the Peutinger Table to have had its left end marked west, its right end east, even though this is not the alignment of the entire map. Certainly Matthew Paris had an idea of consistency of scale (and therewith of direction): on one version of the map is the revealing note 'The whole island should have been longer if only the page had permitted'. At the same time we should not take for granted that he had in mind a map of the sort we are used to; that he would, if he could, have produced that outline of Britain that is familiar to us from maps today. This would be to credit him with cartographic ideas far beyond his time – and even without this his conceptual achievement was a very considerable one. What he produced was essentially an itinerary map of a single route, extended to show what lay on either side and beyond its northern end. The result may not have been very far from a general map, but the idea behind it was quite different.

It is with this same note of caution that we should approach the most elaborate of the itinerary maps from medieval England: the Gough Map, which dates from the mid- or late fourteenth century. Like the Peutinger Table it is named after a former owner, the eighteenth-century antiquary Richard Gough. It is a pity that, unlike Peutinger's manuscript, it has come to have the word map attached to it, for this arouses the wrong expectations, and we are apt to see it as a primitive and faulty geographic map of Britain instead of what it really is, a well executed and efficient itinerary map. But it is a much more advanced production than the maps of Britain by Matthew Paris, more advanced in elaboration and in concept. It takes as its basis not one but almost a network of roads, and in the attention it pays to the relative positions of places that do not lie on the same route we may think that it has begun to justify its name of map. As Mr E. J. S. Parsons has put it, in comparing it with Matthew Paris's maps, 'The characteristic of the Gough Map is the relative accuracy of the topography, to which the itinerary is fitted – not the topography to the itinerary'. At the same time, as Mr Parsons himself shows, it is the roads that provide its framework; in Cornwall, Wales and Scotland,

where there are few roads shown or none, the outline and general topography depart farthest from what our idea of a map would demand. The principal roads on the map radiate from London – five main ones with some branches – but there are also some crossroads, such as from Southampton to Canterbury and from Bristol to Grantham, and in Lincolnshire and Yorkshire a number of local roads radiate from Boston, Lincoln and York. Along each road are numbers giving distances in what were later known as customary miles; the length of this measure varied on no clear regional or other pattern but it was mostly about a mile and a quarter or a little more (about 2 kilometres). The proportional lengths of roads on the map do not correspond to these customary miles, nor to any more regular measure; on the other hand the map-maker clearly had a general sense of scale – roughly speaking the longer the distance on the ground the greater the length on the map – and Mr Parsons has suggested that in some five cases groups of places on the map are in fact correctly placed in relation to each other on a scale of about 1:1,000,000. One such group is Norwich, King's Lynn, Bury St Edmunds and Ipswich, another is Wallingford, Thame, Aylesbury and High Wycombe, and another Arundel, Petworth, Chichester, Havant, Petersfield and Portsmouth. What he implies is that the map-maker made use of local maps, drawn accurately to scale, that were already in existence. This is not impossible, but in the context of the general development of mapping in medieval Europe it seems exceedingly unlikely and in fact the buildings that mark towns on the Gough Map are drawn so large that we have some range of choice in determining their exact positions. But that the possibility should even arise shows how far the compiler had advanced towards the idea of a scale-map. Moreover, although the map's provenance is uncertain there is some reason to think that it originated in a government department; if so, this official use of itineraries in map form would be of great interest, anticipating by nearly a century the use of maps by the Venetian state.

In all the Gough Map is very instructive, for we begin to see what an easy step it is, in both concept and technique, to move from an itinerary map on which distances are given and which shows some sense of the meaning of scale, to a map on which all routes are drawn to a consistent scale; and from this to a general scale-

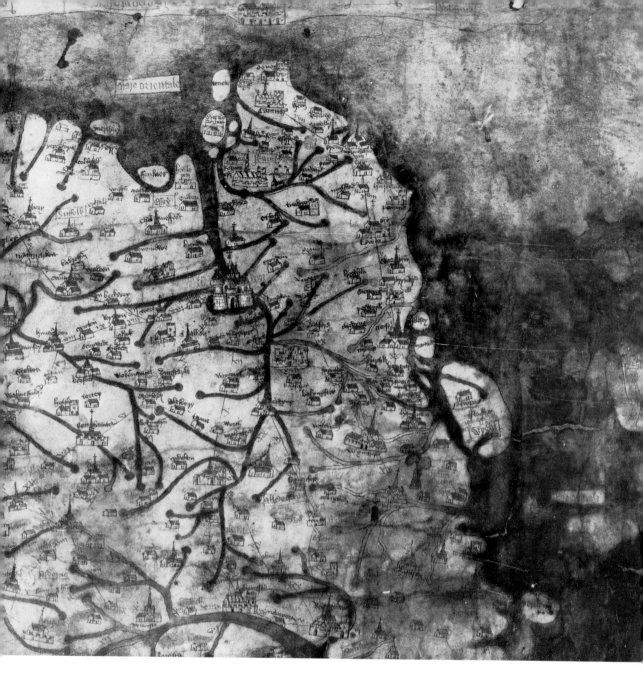

map based on itineraries, the sort of map described by Phei Hsiu, which attains an accurate outline and places all features in their correct relative positions. To achieve this final step, indeed, all that is needed is either a number of cross routes, whose lengths will fix the overall directions of routes radiating from a single point, or an accurate means of measuring the changes of direction along individual routes.

82 South-east England on the Gough Map of Britain, mid- or late 14th century. East is at the top; at the centre are the River Thames and its tributaries, with London marked by walls and towers, and at the bottom is a stretch of the Severn, with Worcester, Gloucester and Bristol. The map is based on the road system, marked by single lines; figures give distances between towns in local customary miles, but the lengths on the map are not proportionate.

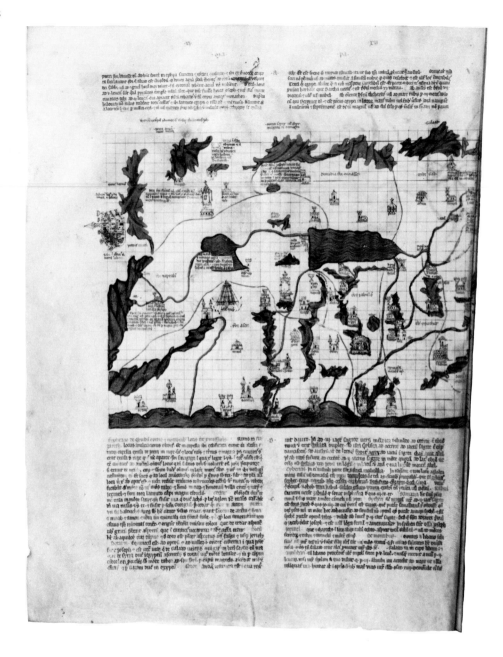

Meanwhile, however, one outstanding production had taken a conceptual leap well beyond the Gough Map. This was the map of the Holy Land prepared by Pietro Vesconte for Marino Sanudo's *Liber secretorum fidelium crucis*, written (as we saw in chapter 3) about 1320. Its most remarkable feature is that it is set out on a rectangular grid. Sanudo explains the principle behind this in the text of his book. The Holy Land, he said, measures 83 leagues from north to south, 28 leagues from east to west:

Let us therefore imagine the Promised Land divided by lines into twenty-eight strips (*spatia*) extending from Mount Lebanon to the desert that leads into Egypt and into eighty-three strips drawn with lines crossing the others from west to east, so that we have a great many squares, each of one league or two miles.

83, 84

144

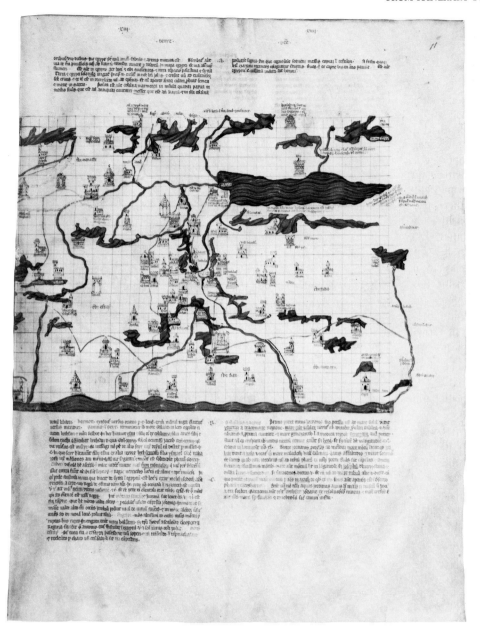

He then proceeds to plot the position of each town in the Holy Land on this grid, proceeding from east to west, strip by strip, and although some places on the borders of the map (such as Baghdad) are badly misplaced he achieves an impressive degree of accuracy. And while the way towns are marked and the precise shapes of rivers and mountains all vary from one to another of the seven surviving manuscripts of

83, 84 The grid of measured squares on this map of the Holy Land, drawn by Pietro Vesconte of Venice about 1320, may conceivably derive from Chinese maps (ills 78, 79); certainly it points to a grasp of cartographic ideas well beyond any contemporary European maps. It was drawn to illustrate a book by the diplomat and propagandist Marino Sanudo, who explains in his text the principles underlying the map.

Sanudo's work, the positions of towns on the grid are always unchanged (apart from the copyists' occasional slips) to accord with this chapter of the text. Sanudo's – or Vesconte's – grid is, of course, a primitive form of the grid of Phei Hsiu and, as Dr Needham has shown, it is possible that the idea came ultimately from Chinese sources by way of Arab writers; one contemporary Arab chronicler illustrates his work with maps of Iran that consist simply of place-names entered on a grid. The chessboard is another possible source of inspiration; the relationship of the chessboard to the topographical map deserves to be explored. Sanudo himself may have seen the method simply as an improvement on that used by Burchardus de Monte Sion who, in his description of the Holy Land written between 1271 and 1291, had defined the positions of places in terms of twelve sectors of a circle centred on Acre. But whatever its source, Sanudo's map was a flash in the pan. It was more than a hundred years before the ideas behind it were again applied to topographical mapping in Europe.

Viewed from this standpoint the fifteenth-century district maps from north Italy are no further advanced than the Gough Map. Most of them, as we have seen, are centred on a particular town, shown by a plan drawn disproportionately large, and this serves as the focus for radiating roads which form the framework for the rest of the map. Distances along the roads are an important feature, for the maps were drawn up, at least partly, for military use. Thus the map of the lands around Parma in 1460 or later marks mileages on the roads themselves; the mid-fifteenth-century map of

IV   the Verona area gives a table of distances beside each of the eight main roads leaving the city; Maggi's map of Padua and its district in 1449 has two tables of distances at the foot. And both the maps of the whole of Lombardy that date from about 1440 have distances marked on the roads. All these maps can be seen as itinerary maps, taking as their basis the roads and the places along them. All show a concern for distances; all pay a general regard to scale. But none is drawn to consistent scale throughout, even if we disregard the central town plans. Probably our earliest Italian itinerary map drawn to scale is the map of the northern boundary of the

kingdom of Naples that was drawn at some time between 1458 and 1516. This is the simplest form of itinerary map, the map of a single route, for the boundary can be seen as a route, like any other line across the landscape; included on the map is a scale of Roman miles.

But if fifteenth-century Italy may or may not have mastered the idea of the scale-map, fifteenth-century Germany certainly had – curiously, perhaps, for in the middle ages there seems to have been less interest in maps in Germany than in Italy and, as we have seen, it was only in south-west Germany that we find a tradition of topographical map-making in the fifteenth century. But it is likely that the earliest scale-maps of any substantial region of medieval Europe were two maps of Germany; likely, but not certain beyond all doubt, for their date and origin are open to question. Both are attributed to Nicholas de Cusa, a theologian and philosopher who died, as Bishop of Bressanone (Brixen) in the Tyrol, in 1464. The two maps are very different, but they may just possibly derive from a common original. One occurs as an addition to the maps of Ptolemy in a manuscript atlas prepared by Henricus Martellus in 1490. The other is a copper engraving that is dated at Eichstätt, in Bavaria, in 1491; but it seems likely that it was left unfinished, and completed only after it came into the hands of Konrad Peutinger, so that it was not actually used for printing until the 1530s. But if the circumstances of the production of Cusa's maps are obscure their antecedents are even more so; behind them, however, lay the interest in theoretical geography among scholars in fifteenth-century Germany and Austria. They are impressively accurate, and as they mark degrees of longitude and latitude astronomical measurement was presumably used to fix relative positions north and south.

But whatever the intellectual antecedents of Cusa's maps they must have had one very practical one in the introduction of the magnetic compass for measuring directions on land. For Cusa's maps are essentially itinerary maps – they were, as one writer has put it, 'the outcome of numerous journeys' with careful measurement of the distances covered. But if the result was to be an accurate map it was necessary to measure not only distances but directions and changes of direction, in accordance with the fifth principle of Phei Hsiu, 'Measuring right angles and acute angles'. It is unlikely that Phei Hsiu can have used the compass for doing this: the earliest reference to the magnetic compass in China (or, indeed, anywhere) is in a book of the late eleventh century by Shen Kua and it is interesting that he suggests strongly that it was being used in surveying for map-making. This was earlier (though only just) than the oldest Chinese scale-maps that survive. We know far too little about the spread of the use of the compass in medieval Europe. In navigation at sea it had reached northern Europe from the Mediterranean by the late fourteenth century, and it seems likely that the fifteenth century saw it increasingly used on land, partly as a means of orienting pocket sundials. It has been said that with the discovery of the compass the direction-map replaced the distance-map; but this is only partly true, for the scale-map, even though based on itineraries, depends on the accurate measurement of angles as much as of lengths.

This is underlined by both the career and the maps of Erhard Etzlaub of Nuremberg: by profession he was a maker of these pocket sundials incorporating compasses (until 1515 when he began to practise as a physician), while two of his printed maps have a tiny dial engraved at the foot and give instructions for using a compass to orient the map. It is, incidentally, interesting that Etzlaub's maps as well as Cusa's are oriented to the south: in the late sixteenth century we can connect a south German tradition of local maps drawn with south at the top with the use of south-pointing compasses there (it is of course mere convention that places the arrow-head on a compass needle at one end or the other). Etzlaub's best-known map covers Germany, the countries of the Alps and the northern half of Italy, and it shows rivers, mountains, towns and roads, these last by dotted lines in which the number of dots gives the distance in miles. Degrees of latitude and a distance scale appear in the margins. It exists in two versions, both published as woodcuts. One has a heading beginning 'Das ist der Rom-Weg . . .' ('This is the road to Rome . . .'), and it has been suggested that it was produced as a guide to pilgrims in the Holy Year 1500, though it may in fact have been published rather earlier, in the 1480s or 1490s. The second

version is dated 1501; the heading begins 'Das sein dy lantstrassen durch das Römisch reych . . .' ('These are the roads through the Roman Empire . . .') and it bears the imprint of the Nuremberg printer Georg Glockenden.

Etzlaub used the same techniques of the scale-map based on itineraries to produce a regional map of the area around Nuremberg. This would be a very natural by-product of his work, whether it was extracted from the itineraries collected for the larger map or simply presented the results of his first surveys. It was a circular map covering a radius from the city of 16 German miles (about 80 statute miles: 130 kilometres), and it was published by Glockenden in 1492. In ill. 86 we see a slightly later map in the same tradition but showing a rather larger area, oriented to the north instead of to the south and with a tiny map of the area around Bamberg, printed separately, stuck on below. On Etzlaub's map rivers and political boundaries are shown, but relief is not, and about a hundred towns and villages appear, marked simply by tiny circles. The map has not a single pictorial feature: more than any other it heralds the arrival in post-Roman Europe of a new cartographical concept, the scale-map.

Overleaf

85 The second of the two woodcut maps of central Europe by Erhard Etzlaub, printed by Georg Glockenden in 1501. South is at the top. Roads are marked by lines of dots, each representing one German mile, as shown by the scale along the bottom; many of the routes radiate from Nuremberg, near the centre of the map, home of both Etzlaub and Glockenden. Where, as in Scotland (bottom right) no roads are shown, the map departs furthest from the truth: it was made by careful measurement of many journeys.

86 Early-16th-century map of the region centred on Nuremberg, similar to one made by Etzlaub and using the same method of showing distance along roads by lines of dots, each equal to one German mile. On this copy of the woodcut another tiny map of the area around Bamberg has been pasted (bottom left); unlike Etzlaub's maps both have north, not south, at the top.

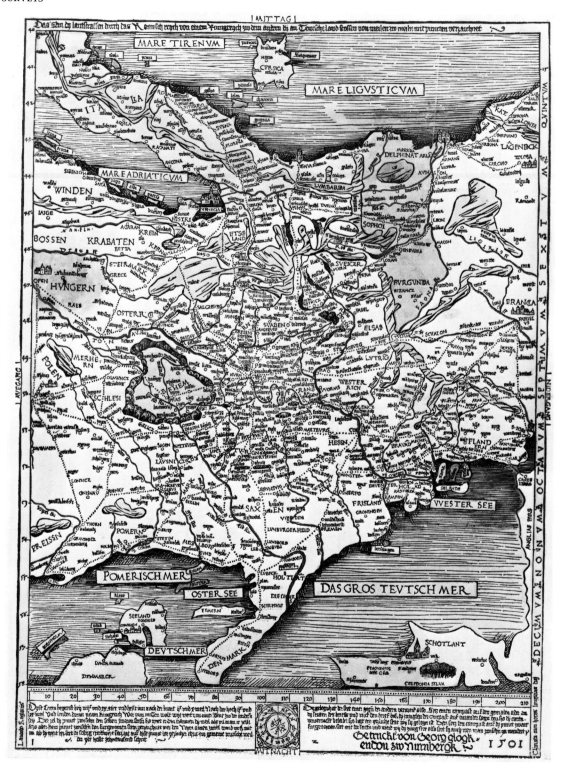

149

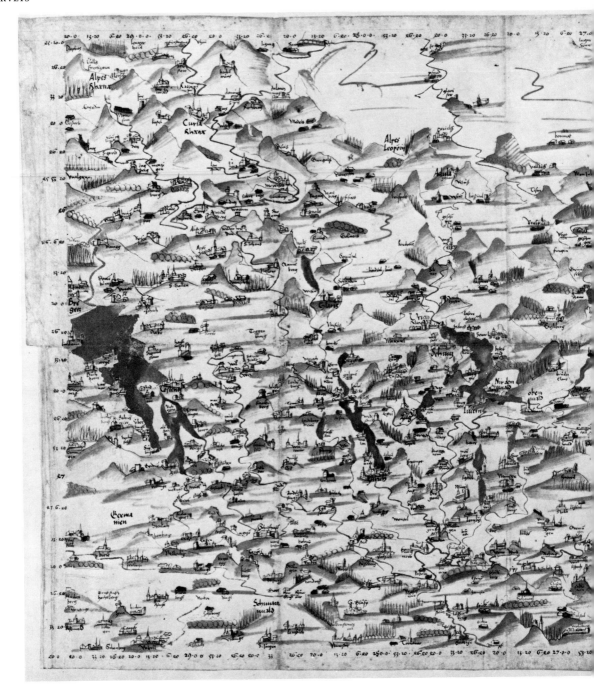

150

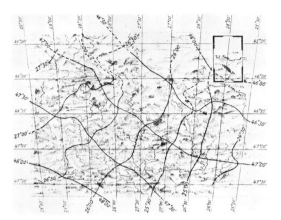

87, 88 Map of Switzerland by Konrad Türst, about 1497. South-east is at the top; Lake Constance is at the centre of the left side, Aosta is in the top right corner. The tiny sketches of towns and villages are drawn from life. Lines of latitude and longitude, drawn as if taken from the map (above), show that it is not entirely accurate; even so, it was a great achievement for its time.

But a truer precursor of immediate developments in the sixteenth century was the map of Switzerland by Konrad Türst. Türst, like Etzlaub, was a physician, but of very much higher standing, for in this capacity he served the Emperor Maximilian I. Between 1495 and 1497 he wrote *De situ Confoederatorum descriptio* ('Account of the country of the Confederation'), a description of Switzerland that was not published in his lifetime; it is known from four manuscripts, of which two have his map attached. Like the maps of Cusa and Etzlaub this was one of the great achievements of European map-making in the fifteenth century: a map drawn true to scale from itineraries made in a country where the mountains posed very special difficulties of access and measurement. And like the maps of Cusa and Etzlaub it is oriented to the south; it also has degrees of latitude and longitude in the margins. Reproduced here is one of the two copies of Türst's map, and beside it a diagram of lines of latitude and longitude as they would appear if laid in their true positions on the basis of the map; they look very shaky, very far from the firm grid of straight lines that would appear if the map were completely accurate — but given that the

87, 88

151

regional map drawn to scale was an innovation, given the peculiar difficulties of the Swiss terrain, the diagram is in fact a striking tribute to Türst's skill as a cartographer and to the success of his methods.

But if Türst's map has clear affinities with the maps of Cusa and Etzlaub it is no less clearly connected with the regional maps of north Italy. Even its orientation may derive from Italian maps rather than from the German tradition, for the picture-maps of north Italy are oriented to the north so that the Alps form a natural horizon and in Switzerland the same principle would produce a map with south at the top. And Türst's map, while drawn to scale, is in the style of a picture-map. Settlements are represented by thumb-nail sketches, not in con-ventionalized form but showing the actual appearance of each place; in a few cases where a name has been accidentally omitted the village can be identified from the sketch alone. The sketches must come from drawings made on the spot and they confirm that the map is based on a series of measured itineraries made for the purpose. Türst had connections with north Italy: he visited the court of the Duke of Milan in 1493 and again in 1497 and, significantly, he sent a copy of his map there when it was completed. In the work of Türst we see the welding together of the picture-map and the itinerary-based scale-map, the union that was to produce the map of the Black Forest by Tibianus 1 that we saw in the introduction and innumerable other maps in sixteenth-century Europe.

10

Sixteenth-century Europe

IN A sense we have now completed the task set in the introduction. We have seen what course of development lay behind the topographical maps that were being drawn to scale in sixteenth-century Europe: maps like that of the Black Forest by Tibianus in 1578. We have seen how they stem from the various traditions of picture-maps to be found in late-medieval Europe, and how these picture-maps can be paralleled in other societies, in different ages and different continents, that had reached a comparable stage of cultural and artistic development. We have seen how three earlier societies – Babylonia, Rome and China – had succeeded in making the same transition from picture-map to scale-map that occurred in Renaissance Europe, but without permanently establishing this new sort of cartography. We have seen how behind the picture-map lies a yet more primitive form of mapping in the symbol-map that we find among peoples who were mostly less advanced in culture than those who made picture-maps. And we have seen how the capacity to make maps, the habit of mind that thinks cartographically, is something that cannot be taken for granted even in quite advanced societies; it involves ideas that are far less simple, far less a normal part of human intelligence than we are apt to suppose, and if we have interpreted the evidence correctly map-making seems to have appeared rather erratically in the course of man's history.

But to leave the story there would be to leave it unfinished. We should at least follow into the sixteenth century the lines of development that we have identified in the middle ages so as to see how they interacted to produce a firmly established tradition of scale-maps based on measured survey. And we shall trace the continuing threads that link the picture-maps of the middle ages with the pictorial elements in

topographical maps of today, such as the maps of the areas around Canterbury and Zermatt that we looked at in the introduction. These are the themes of our two concluding chapters. In some ways they differ from the rest of the book. So far we have been looking at areas of cartographical history that have been little explored and where scholarly investigation has concentrated on particular maps rather than on general patterns of development. Consequently every chapter has contained suggestions, hypotheses and unanswered questions that can be tested or resolved only by further research. The history of maps in the European tradition from the sixteenth century on, however, is very much better known; while we must have more to learn from future research its main lines of development and principal landmarks are established beyond reasonable doubt. What is said in these two chapters, therefore, is far more firmly based than what has gone before. Whereas so far we have been offering a comprehensive history of the early stages of map-making, taking into account every scrap of the very scanty evidence available, we are now following only the one or two strands of development that relate to our theme and ignoring many important aspects of the subject. The number of maps surviving from the sixteenth century on is enormous, and whereas the maps illustrated or discussed in detail so far have practically chosen themselves from the very few that exist, those in this chapter and the next have been selected from hundreds or thousands of possible examples. That this is so is itself one of the changes in the sixteenth century that we shall look at.

But as a first introduction to the topographical maps of the sixteenth century we cannot do better than look at the maps by

Leonardo da Vinci, for they draw together and illustrate many of the conceptual and technical difficulties that fifteenth-century map-makers were grappling with. We may even have met his work already, for it has been suggested that his ideas lay behind the detailed bird's-eye views of cities that came from Rosselli's workshop in late-fifteenth-century Florence. It is no more than one would expect that Da Vinci, artist and natural philosopher, would take an interest in problems that involved several of his activities so closely and in his maps we see one of the best minds of his age brought to bear on them. The most important of Da Vinci's surviving maps are the eighteen in the Royal Collection at Windsor Castle, from which the two reproduced here are taken. They date from the earliest years of the sixteenth century. Some, like ill. VIII, which shows part of an embanking scheme to divert the River Arno, near Florence, stem from his interest in hydraulics and the flow of water; in 1500 he had written a treatise 'On the movement and measurement of water'. Others, like ill. 89, come from his work as a military engineer; in 1502 he entered the service of Cesare Borgia and saw something of the war of 1503 between Florence and Pisa. But in them all we see the eye of the artist, interested in the problem of transferring the features of landscape on to paper and experimenting with techniques which, partly with hindsight, we would call cartographic.

Certainly in two of his maps we see the culmination of the medieval Italian tradition – if tradition there was – of outline plans of towns drawn to scale. A sketch plan of the main streets and gates of Milan recalls the description of the fourteenth-century map of Florence by Da Barberino (chapter 4), for there is a note of the distance between each gate and at one side an addition of these figures to give the total length of the city walls; below it is a bird's-eye view of the city showing the principal buildings. And 89 Da Vinci's plan of Imola, while recalling the town plans of the north Italian district maps in its notes of the distances to neighbouring towns, differs from them in its great detail, showing even individual houses, and in being drawn strictly in outline; even such obvious candidates for pictorial treatment as the cathedral and the castle are represented simply by their ground-plans. The same plan of Imola gives us

a hint of Da Vinci's concern for direction: from the city centre eight lines marking the principal compass points radiate to the map's circular frame, while fainter lines mark eight subdivisions within each sector, so that the whole circle of the compass is divided into sixty-four segments. And on each of two maps of areas in south-east Tuscany Arezzo is made the centre of a small pattern of compass lines, twelve in all, so that each quadrant is divided into three. In his landscape maps Da Vinci occasionally used outline plans where medieval practice would have had a picture. We see this in ill. VIII, where the works of the embanking scheme are shown in plan; again on a related sketch, showing these same works in a wider setting, mills appear in plan, not pictorially, and two pen-and-ink sketches of a system of streams near the village of Vincio use outline rectangles to mark houses.

But most of Da Vinci's landscape maps, unlike his town plans, are basically pictorial, and we can see how he was aware of the problem of reconciling this approach with the cartographic concept of representation to a uniform scale. Several of his maps have a specific note of scale in the form of a series of dots or dots and bars, numbered or unnumbered, in one margin. The assumption is that these maps, while pictorial, are drawn not in perspective but in a single scale throughout and this is borne out by measurement, though his scales do not seem to have achieved total accuracy or total consistency. On one map of the Tuscan coast from Lucca to Campiglia with the immediate hinterland, drawn probably for the war of 1503, landscape is shown in a multiplicity of bird's-eye views, each with its own horizon; this was a technique that lent itself very well to uniformity of scale (as we see on the map of the Black Forest by Tibianus), but here, conversely, an element of overall perspective is introduced by showing the more distant places smaller than the nearer ones. Da Vinci was aware too of the problem of showing relief – the problem that more than any other bedevilled the move away from pictures in mapping in the sixteenth century and later. On the map of the Tuscan coast he shows hills in exaggerated outline; on others, such as those of south-east Tuscany and of the area around Vincio, he uses various types of shading on hills. On one map showing the whole river system of the Arno and its tributaries, we

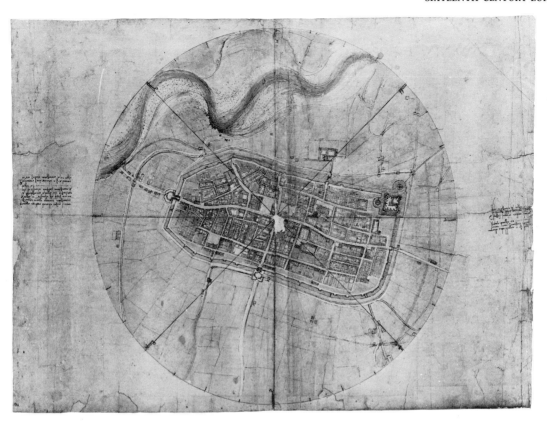

find in some places what looks like an experiment in contour shading, showing the high hills dark, the lower ones lighter, but other hills on the same map are marked by the normal shading that simply places one side in shadow. And in marking settlements Da Vinci uses all three methods of the late fifteenth century: tiny circles, conventionalized views of buildings or town walls, and actual thumb-nail sketches of the places shown. But throughout all this interest in techniques, all these experiments in cartographic representation, we keep seeing the hand of the great artist: we see it in ill.VIII in the extraordinarily skilful and powerful painting of the currents and eddies in the river – it was the natural philosopher, the scientist, who observed and measured them, the artist who set them down on paper.

In many ways Da Vinci anticipated the developments in topographical mapping that occurred throughout Europe in the sixteenth century: in his concern for scale, his move away from pictorial representation towards outline plan, the union in himself of the surveyor and

89 As artist and military engineer Leonardo da Vinci had a lively interest in cartography. His plan of Imola is the earliest undoubted scale-map of an Italian city since the Roman period. The notes on either side are in mirror-image writing.

military engineer with the artist. But most of all he typifies what was to come, quite simply in his interest in the problems of cartography and his use of maps to set out and clarify all sorts of problems involving landscape: drainage, military strategy and tactics and so on. For what above all distinguishes the cartography of the sixteenth century from that of the fifteenth is its sheer quantity. In medieval Europe we found odd pockets of map-making, the occasional individual who drew maps, the strictly limited local or regional tradition; but these stood out against the background of a society that knew little of maps and simply did not think in cartographic terms when confronted with the need to record or communicate topographical information, whether it concerned half a field or half a continent. Suddenly this changed: it was as though the flood-gates were opened at the turn of the century, and the value and use of maps were quickly recognized throughout southern, western and central Europe. We do not have to search diligently to find sixteenth-century maps from Italy, the Iberian Peninsula, France, Germany, the Netherlands or England: they survive in their hundreds.

This map-consciousness spread with great speed, right from the start of the century. In 1505 we find the Emperor Maximilian I commissioning Johann Stabius of Steyr to travel throughout the lands of Austria and gather materials for a map, and it may well be that the impetus to the spread of map-making came from the use of maps for official purposes; that the rulers, statesmen and generals of the sixteenth century realized how valuable maps could be for administration and, particularly, warfare, just as the Venetians had done a generation or so earlier. We have collections of maps of English coasts and fortresses made under King Henry VIII in the 1530s and certainly by the second half of the century maps were regarded as an essential tool of government. An inventory of 1575 lists some 250 maps that belonged to Viglius van Aytta, President of the Council of State for the Spanish Netherlands. And in England in 1592, in 'A treatise of the office of a councellor and principal secretarie to her Majestie', Robert Beale, Clerk of Queen Elizabeth I's Privy Council, wrote that

A Secretarie must likewise have the booke of Ortelius' Mapps, a booke of the Mappes of England . . . and if anie other plotts or mapps come to his handes, let them be kept safelie.

These were precepts that were certainly put into practice by Lord Burghley, who was Secretary of State and then Lord Treasurer throughout nearly the whole of Elizabeth's reign. There survives a large collection of manuscript and printed maps that belonged to him, and many of them, besides others in the government records, have notes and endorsements in his own distinctive hand. It was not just maps of entire countries and regions that were used for the work of government: Burghley's collection includes maps of particular districts and places. And, although the work of government may have stimulated the spread of mapping, maps in

sixteenth-century Europe were not used only for official or military purposes: we find maps performing almost as many functions as we do today.

At the same time we should not exaggerate Europe's new-found map-consciousness. We occasionally find a surprising reluctance to think in terms of maps. Richard Hakluyt is a case in point. His two works, *Divers Voyages* (1582) and *Principall Navigations* (first published 1589 and reissued in three volumes in 1598–1600), present collected accounts of voyages of discovery and form an outstanding source of information about the explorations of his age. Here, if any, were works that called for illustration by detailed maps; but those in the publications of 1582 and 1589 are quite inadequate, and Hakluyt's writings suggest that he would not automatically turn to maps as many of his contemporaries had learned to do. Again, there were geographical limits to the use of maps in the sixteenth century: in northern and eastern Europe the idea spread much more slowly. There is only a single sixteenth-century map in the Scottish Record Office, though plans and views of Scottish fortresses and towns, among them the earliest known plan of Edinburgh, were made by the English in the successive invasions of the 1540s. The earliest known maps of the provinces of Sweden date from the second and third decades of the seventeenth century.

Nor, when maps came into general use in sixteenth-century Europe, did they necessarily represent the most advanced cartographical techniques of the time. Etzlaub, Türst and Da Vinci were all producing scale-maps made from measured surveys at the start of the century; but as topographical maps became more widespread what we see is a great blossoming of the picture-map. Ill IX is an example that must serve to represent a vast number of such maps: a page from a manuscript book of 1543 showing, in perspective, the course of part of the River Vilaine in Brittany, connected probably with a royal order of 1539

90 The area around Rennes on a picture-map of 1543. It comes from a fourteen-page manuscript atlas of the River Vilaine; concerned primarily with navigation, it pays particular attention to the river's locks and (as here) bridges.

to have the river made navigable to large boats. We could say, indeed, that the picture-maps that we have from medieval Europe are no more than casual precursors of the real tradition of picture-maps that came into being only in the sixteenth century, a time when the picture-map in Europe was already competing with the more advanced scale-map based on survey. Moreover, this tradition was a strong one and was still flourishing at the end of the century. Thus by the 1580s and 1590s scale-maps of estates were being produced in England but had by no means replaced picture-maps made without measurement; Ralph Agas's map of Toddington in Bedfordshire, drawn in 1581, a carefully surveyed scale-map, can be matched by, for instance, maps of Brill in Buckinghamshire made for New College, Oxford, in 1590, which are carefully drawn and ornamented but owe nothing to measurement on the ground and cannot be used for accurate indications of either direction or distance. In Franconia there was a recognizable tradition of picture-maps, among them works of the court painters at Würzburg, Speyer and Coburg (ill. 100 is an example); some were drawn – like the fifteenth-century picture-maps from France and the Netherlands – for production in court in cases over village boundaries, hunting rights and so on. And the various techniques of the picture-map that we have seen in earlier chapters can all be paralleled in the sixteenth century. Thus a map of Dartmoor drawn about 1540 in connection with pasture rights is an excellent example of the sort of diagram-map discussed in chapter 5. Again we find, for instance on a picture-map of the Fils valley in Württemberg, drawn about 1534, the same method of opening up a landscape so that we have in effect two picture-maps facing each other with opposing horizons, a technique that we have seen on an Assyrian bas-relief (chapter 3), the Madaba mosaic and the Burgundy boundary map of 1460.

Nor were maps of rural landscape – estate maps and the like – the only picture-maps to come into their own in the sixteenth century. The detailed bird's-eye views of towns produced in late-fifteenth-century Italy came to be widely copied. Even De' Barbari's magnificent plan of Venice in 1500 was only the first of a number of equally large, detailed and accurate city views; those of Augsburg by Jörg Seld in

1521 and of Antwerp by Virgilius van Bologne and Cornelius Grapheus in 1565 are outstanding examples, but there are a number of others, among them the earliest map of London, drawn in the 1550s and known only from the chance survival of two of probably twenty engraved plates that were used to print it – no printed copy has survived. And there were many other bird's-eye views of towns that were less ambitious: the method became a recognized and well developed form of topographical art and between 1572 and 1618, in their *Civitates orbis terrarum* ('Cities of the world'), Georg Braun and Frans Hogenberg published six volumes of engraved views of towns throughout the known world.

It is, then, against this background that we should see the development and spread, in sixteenth-century Europe, of the topographical map drawn to scale from measured survey. Its way was being prepared by the picture-map, which spread the idea of cartographical representation. But in the case of regional maps, maps of entire provinces, it was often the scale-map that led the way. In south-west Germany the small-scale maps of Cusa and Etzlaub were followed by Martin Waldseemüller's of the Rhine from Basle to Bingen in 1513 and one, drawn by Sebastian Münster a few years later but left unpublished, showing the Rhine from Basle to a point a little below Cologne. It was Münster who in 1525 remarked that a failing of maps of Germany was that they often treated the 'miles' of colloquial usage, which varied from one place to another, as though they were standard units of length. Elsewhere too we find that regional maps were being drawn to scale in the first half of the sixteenth century. Gerolarmo Bellarmato, in the dedication on his map of Tuscany published in 1536, tells how it was based on notes and measurements made from his own travels through the area on horseback; his map is of particular interest in that he tries to distinguish the sizes of settlements in terms of

91 Part of a map of the village and fields of Toddington, Bedfordshire, by Ralph Agas, 1581. Its competence and accuracy in setting out the intricacies of open-field strips show how far the 16th century had advanced in cartographic concepts and surveying techniques.

the number of hearths they contain. In the Netherlands the map of Brabant by Jacob van Deventer, the earliest to survive of his notable series of maps of the provinces, dates from 1536. Here England lagged far behind: the first set of county maps of England and Wales, drawn to scale by Christopher Saxton, was not produced until the 1570s.

All these are well-known maps and the way that scale-maps of individual regions developed and spread in sixteenth-century Europe has been thoroughly explored by historians of cartography. This is not the case when we turn to scale-maps of smaller areas: maps of particular towns, estates and so on. Yet certainly in England, probably in many parts of Europe, these were the earliest type of map to be drawn to scale. Surprisingly, no one has ever systematically investigated when, where or by whom the first scale-map was drawn in England, in the Netherlands, in France, in the Iberian Peninsula or in eastern Europe – nor the first map of a small area drawn to scale in Germany. This is partly because the distinction between the medieval picture-map and the scale-map of later periods has not been fully appreciated: it would be assumed that the idea of uniform scale lay behind any map, so that divergence from it in the middle ages resulted from inadequate techniques rather than from difference in intention. The appearance of a topographical map drawn accurately to scale would thus reflect technical advance but not a radical change in cartographic thinking.

We have seen how at the end of the fifteenth century Erhard Etzlaub of Nuremberg produced scale-maps of the area around that city and of routes over much of central Europe. In 1507 he was commissioned to make a measured survey of the estate of Hauseck that had been bought by the city of Nuremberg; we do not know whether he produced a map from this survey, but if he did it seems at least likely that it will have been drawn to scale. In the Netherlands we find that local maps drawn to scale in the sixteenth century were referred to as maps using 'the little measure' or 'the little foot', meaning the tiny length on the map that corresponded to the full-length foot or other measure on the ground – just such a phrase as might occur to people to whom the idea of a scale-map was a new or unusual one. Thus in

1537 in a dispute between the Emperor Charles V and Jan van der Wouwer about fishing rights in the Biesbosch (the area east of Dordrecht, between the rivers Maas and Waal) Jan van Barry was ordered to survey and map the fishing bounds, using the little foot – we notice in passing that using maps in law-suits was a continuing tradition in the Netherlands. And in 1539 Jacob van Deventer agreed to make 'a map with the little foot' ('een Carte opte cleyne voet') of Delfland; other references in the contract to the little foot include the requirement of marking churches 'using the little foot in accordance with the proportion of the large foot'. The usage continued in Dutch until at least the end of the sixteenth century.

What must be one of the first scale-maps from England appears in ill. 92; it shows the town of Portsmouth on the south coast and dates from 1545. It has no scale-bar, but along the bottom of the map is written 'Thys plat ys In every Inche on C fote', that is a scale of 1:1200. The map's purpose was to show how the town's fortifications could be improved and extended; further proposals, probably made a year later, were sketched on afterwards in pencil. It was drawn up following French attacks in the area in the summer of 1545 and a visit to the town by King Henry VIII. The early maps of Portsmouth and its surroundings are the subject of an important study by Mr D. Hodson. His work has revealed an extraordinary number of large-scale maps from this area: over 350 dating from before 1801, of which twenty-two come from the sixteenth century. By 1800 in fact, Portsmouth must have been mapped better – or at least more often – than almost any place of comparable size anywhere in the world. The reason is not hard to find: from the mid-sixteenth century Portsmouth was built up first into an important fortress, then into one of Britain's principal naval harbours and dockyards, and almost every one of its maps was drawn for a naval or military purpose. This reflects the part played everywhere by military

92 In England military engineers were among the first to draw maps true to scale. This very early example shows the new fortifications planned for Portsmouth in 1545, after the French attack and landing on the Isle of Wight.

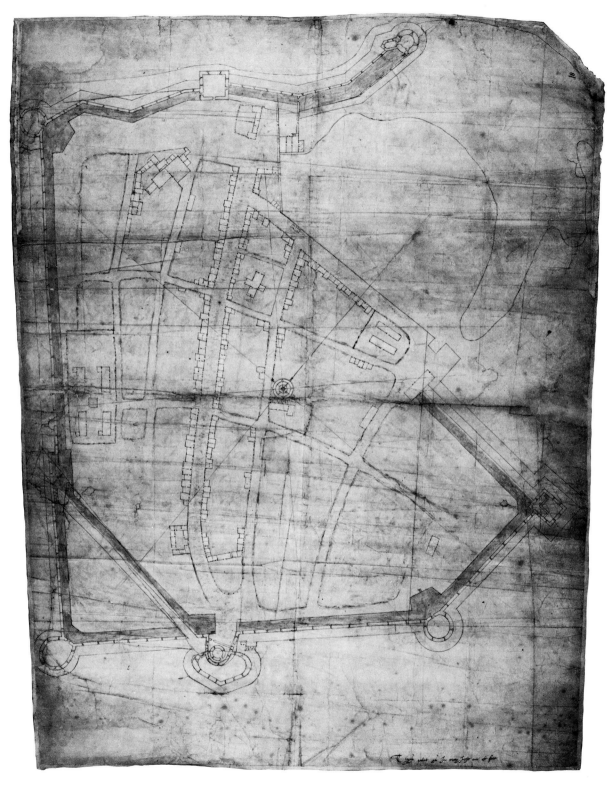

needs in spreading the use of maps in the sixteenth century and later, as foreshadowed by the fifteenth-century district maps from north Italy and by the maps that Leonardo da Vinci drew as a military engineer. Moreover, of the twenty-two sixteenth-century maps of Portsmouth only one is not drawn to a consistent scale, and this is the earliest of all, dating from about 1539. It seems very likely that military engineers did a great deal to develop the use of consistent scale on maps of small areas. On the maps that belonged to Lord Burghley we commonly find a note of the scale on those drawn by military engineers from the late 1540s onwards, but only from the 1570s on the other local and regional maps. After all, precise measurement, as well as the exact details of topography, was essential for the engineer who was to construct the fortifications that were planned. This in turn, like the modest spread of local map-making in fifteenth-century Europe, may have derived from developments in the plans used by all architects and masons. But this is no more than a guess.

The sixteenth century not only saw the spread of topographical maps in general and in particular of scale-maps drawn from survey; it also saw the introduction of a new technique for making scale-maps – triangulation. This involved the very precise measurement of a single base-line and of a great many angles to construct a framework of points, accurately defined in relation to one another, within which the detailed features of the area could be fitted. As we have seen (chapter 9), this was almost the exact opposite of the careful measurement of many itineraries that Phei Hsiu and the cartographers of early-sixteenth-century Europe took as the basis for their scale-maps. These built up the map of the entire area, from its centre to its outer limits, by measuring routes within it. Triangulation, on the other hand, started by fixing points at the map's limits and elsewhere, and only then filled in lines of routes and other details. It had the great advantage of ensuring that the map's overall proportions were correct, whereas the itinerary-based map could contain cumulative errors that would produce distortion. Again, the accuracy of the itinerary-based map could be checked only by astronomical observation, which in practice could determine only how far north or south particular points

were and could be applied only to maps of large areas. Triangulation, on the other hand, had built into it a check on its own accuracy in its requirement of angle-measurements that would bring its series of triangles back exactly to their starting-point on the measured base-line; this would operate just as much over a small area as over a large one and moreover an independent check could always be introduced by seeing whether the calculated length of a side of one of the triangles really corresponded to its length on the ground. The only disadvantages of triangulation were the more or less complicated calculations it involved and, above all, the difficulty of measuring angles with the necessary degree of accuracy. This last was crucial. Works on trigonometry, the basis of triangulation, were known to scholars in the medieval west from the twelfth century onwards, but their application to map-making, even when the idea of maps and of scale-maps had become widely accepted, depended on the development of increasingly precise instruments for measuring angles. These were seldom satisfactory for any but elaborate surveys made by highly skilled map-makers; even in the eighteenth century we find writers on estate surveying – such as John Gray in 1737, Arthur Burns in 1776 – advocating surveys by methods that did not involve measuring angles, as being simpler and less liable to error. Their comments underline the difficulty of angle measurement and help to explain why triangulation was not generally adopted more quickly.

For the technique in fact came into use very slowly. One of its greatest early achievements was the map of Bavaria by Philipp Apianus; based on a survey made in 1554–61, it measured some 21 feet square (6·5 metres). This map is lost, but a reduced woodcut version was published in atlas form at Ingolstadt in 1568; put together, the twenty-four leaves in the book form a map some 5 feet square (1·7 metres). But although Sebastian Münster, who died in 1552, had proposed a map based on the triangulation of the whole of Germany he was well in advance of his time: triangulation was first used for a map of Baden about 1600, for Schleswig-Holstein not until the nineteenth century. It was not until the work of Jacques Cassini in 1739–44 that triangulation was applied to as large an area as the whole of France, and the linking of the

93

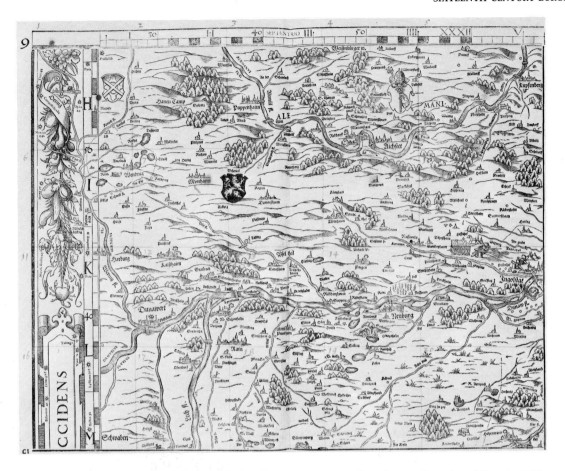

British and French systems of triangulation in 1790 was an important achievement in the scientific cartography of its age.

In view of this it is not surprising to learn that Christopher Saxton's maps of the English and Welsh counties in the 1570s were probably simply itinerary maps, made without triangulation. So too was the map of Denmark made on the orders of King Christian III and published in 1552, even though it was produced by a Copenhagen professor of mathematics, Marcus Jordanus. Perhaps the first map from Denmark to use triangulation was of the island of Hven in 1596. Its author was the mathematician and astronomer Tycho Brahé, who had his observatory on the island and who published the map in a work entitled *Epistolarum astronomicarum libri* ('Books of letters on astronomy'). He tells in the text how he determined astronomically the exact position of the centre of the island and gave the latitude and longitude

93 A page from the woodcut atlas of Bavaria, 1568, that derived from the survey of Philipp Apianus in 1554–61, one of the most notable achievements of 16th-century mapping. We see the upper Danube from Donauwörth to Ingolstadt; north-west of Ingolstadt soldiers and cannon mark the battle of 1546.

on this basis, but he does not say how the map was drawn. From other records, however, it can be shown that he used triangulation, and this is indeed what we would expect from his keen interest in instruments for measurement. There was always a close link between the academic mathematicians and the application of the technique of triangulation even to the humdrum and lowly business of mapping very small areas; it is an unusually early case of practical techniques developed from abstract theory. The practical use of triangulation is believed to have

163

been started by a book by Gemma Frisius (Jemme de Fries) published in 1533. Five years earlier Sebastian Münster had published a work, in German, which took the area around Heidelberg as an example to show how a map could be constructed by using the land-compass. Now Frisius, in his *Libellus de locorum describendorum ratione* ('Little book on the way to depict places'), published both in Latin and in Dutch, showed how triangulation could be used; his example was an area in north Belgium, taking the distance from Antwerp to Brussels as his base-line. He was himself a mathematician, at Louvain (he matriculated in 1526); significantly, among his predecessors there was Jacob van Deventer (matriculated 1520) and among his pupils Gerardus Mercator who not only produced the geographic maps by which he is best known, but also worked as a surveyor, developed his own instruments and made a map of Flanders in four sheets in 1537–40.

It is not at all clear how far triangulation was used for large-scale mapping in the sixteenth century. It was only one of the methods available for making maps to a consistent scale, and it seems unlikely that its introduction actually led the way in the spread of the scale-map. Mercator probably did not base his map of Flanders on triangulation, though it has been suggested that he may have used it to check his results. With the rapid development of instruments it is far more likely that Deventer used triangulation for the work that occupied the last seventeen years of his life: a collection of plans of all the walled towns in the Netherlands, some 250 in all, drawn to a uniform scale (ill. 94 is an example). He undertook this gigantic task in 1558 under a commission from King Philip II; it was supposed to be completed in two years, but it was 1572 before the drafting was finished, and he was still working on the plans when he died three years later. Certainly before Deventer even began this work triangulation had been used in 1547 for a map of Vienna by August Hirschvogel who, perhaps significantly, came from Nuremberg. Some clue to the spread of triangulation is provided by the increasing number of treatises published in the wake of Frisius, giving instruction in the method. Thus in Switzerland Sebastian Schmid's book appeared in 1566: *Underrichtung, wie man recht und kunstlich ein jede Landschaft abcontrefehen* *und in Grund legen solle* ('Directions for portraying and setting down an area of country accurately and elegantly'). This described three methods of drawing maps to scale: measuring distances from a single centre, triangulation, and combining measurements of direction and distance. In England a work by William Bourne, *A Booke Called the Treasure for Traveilers*, published in 1578, gave an account of triangulation abridged from a more extended treatise which survives in manuscript and which illustrates the method by applying it to an area in Kent and Essex. But at present we can do little more than guess how far triangulation was used for scale-maps of towns, districts or estates in the second half of the sixteenth century.

It was the gradual spread of the scale-map, irrespective of how it was constructed, that lay behind another important change in topographical mapping in sixteenth-century Europe: at the beginning of the century mapping was the work of the artist, by the end of the century it had passed into the hands of the surveyor. The outward effects of the change can be illustrated neatly from two plans of Rome published in the 1550s. The first, the woodcut plan of 1551 by Leonardo Bufalini, is the earliest plan of Rome to be drawn true to scale throughout, and it is entirely surveyor's work, simply presenting in graphic form the results of accurate measurement: buildings are shown by their ground-plan alone – the eye and hand of the artist have played no part in the plan's construction. The engraving by Francesco Paciotti on the other hand, though published six years later, belongs to the older style of mapping; it is entirely a bird's-eye view, the work of an artist just as much as De' Barbari's map of Venice in 1500.

The process of change must have varied from one country to another, but it can be illustrated from what happened in the Netherlands. The local maps that were drawn there in the early sixteenth century were the work of men who are otherwise known as artists or who are specifically called painters. Willem Croock, whose surviving maps of Amstelland and the North Quarter of Holland date from the 1520s, had in 1520 painted the standards for the warships of the Emperor Charles V; Jan de Pape produced a map in 1527 for a lawsuit between

95, 96

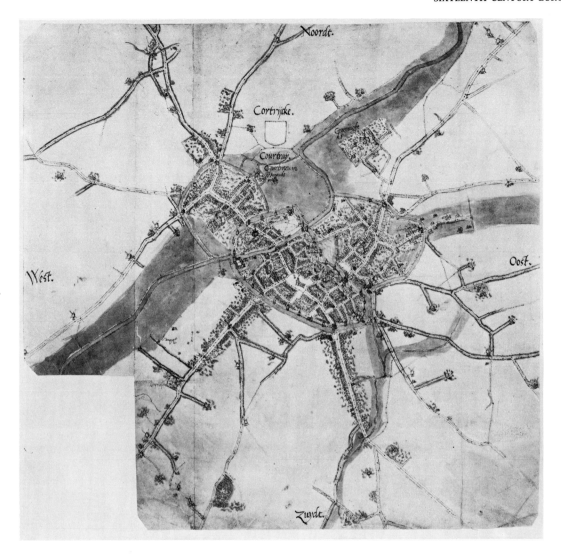

94 Plan of Courtrai, one of 250 plans of towns in the Netherlands drawn by Jacob van Deventer between 1558 and 1572, showing the minute detail of his work. After his death in 1575 the collection remained in manuscript, unpublished.

the Emperor and the town of Zierikzee, while a year or so earlier he had been employed to paint the imperial arms above the fireplace in the exchequer office at The Hague. Then we sometimes find the artist producing the map from the work of the surveyor: Pieter Gerritszoon, working at Haarlem, had the help of a surveyor (*geometricus*), while the first map to be published by the great Antwerp printer, Christophe Plantin, in 1556, was one of the Vermandois that had been drawn and engraved by the royal painter Antoon van den Wijngaerde, from a survey by Jan Surhon. But from the middle of the century it was usually simply a surveyor who made the map; already in 1540 Deventer

was being called just 'surveyor' (*landmeter*), and in 1598 it was the surveyor Simon van Buningen who produced a set of sixty-eight maps of the estates of the former abbey of Rijnsburg. This did not mean that the maps were necessarily less pictorial. Looking at the published collection of maps that have come to light

165

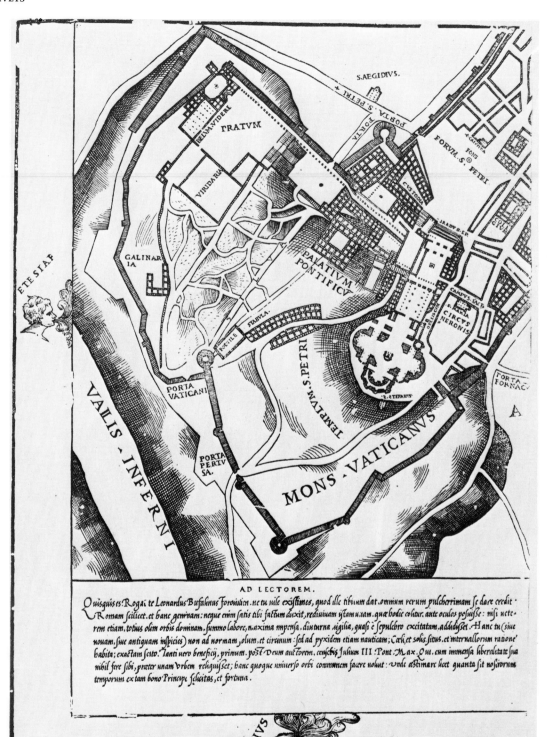

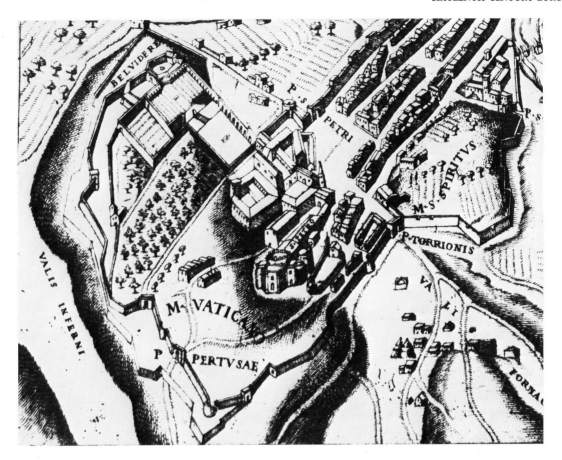

95, 96 St Peter's and the Vatican Palace as shown on two mid-16th-century printed maps of Rome. One (left) is by Leonardo Bufalini, 1551: the earliest scale-plan of the city since classical antiquity, showing features by their outline ground-plans. The other (above), by Francesco Paciotti, 1556, is a picture-map, a more or less realistic bird's-eye view in the style developed in Italy at the end of the previous century (as in ills 34, 38, 39). Each method gives some information that the other does not: we see here the start of the separate traditions of maps and bird's-eye views. On Paciotti's map we see that St Peter's lacks its dome: Michelangelo's design had not yet been put into execution.

among records of legal suits at Mechelen, we find that the three earliest specifically by a surveyor ('gezworen landmeter') date from the late 1540s; they all include a strong pictorial element, showing houses, trees, windmills and other miniature views in multiple perspective.

And even in 1585 it was the painters Ferdinand Bart and Pieter Clays who made a map of the district around Ostend and Sluis for the commander of the Spanish forces in the area. But the change of emphasis in the course of the century is clear and behind it lie developments of great significance in the history of topographical mapping.

Moreover there was a corollary to this. It was not just that the man of 1600 who wanted to have a map made would send for a surveyor, whereas his great-grandfather would have looked for a professional artist. After all, as we have already seen (chapter 5), by the sixteenth century surveyors were a long-established profession in the Netherlands, their function being to measure properties, define boundaries and record their findings in written descriptions. But now making maps did not become merely a useful adjunct to their craft: it began to be an essential part of it, so that it came to be assumed that the end-product of a survey would normally include a map and might indeed consist of nothing else. In England this development occurred towards the end of the sixteenth century. In 1577 Valentine Leigh, in a book called *Surveying of Landes*, advises the surveyor to accompany his completed survey with a map; but his treatise otherwise lies in an older tradition, and he does not explain how the map is to be made. Certainly one has no difficulty in finding estate surveys of this time that are drawn up on the old pattern: elaborate descriptions of the size and position of each piece of land, unaccompanied by a map. An example is the series of nineteen surveys of estates, mostly in Norfolk, belonging to King's College, Cambridge, drawn up between 1564 and 1587. But at Oxford in the 1580s and 1590s All Souls College was having surveys of its lands embodied in a set of superbly detailed maps, and in 1607 John Norden in his treatise *The Surveyor's Dialogue* was including 'how to take the plot of a mannor' as a normal part of a surveyor's expertise. His contemporary Ralph Agas expressed himself more forcibly:

No man may arrogate to himselfe the name and title of a perfect and absolute Surveior of Castles, Manners, Lands, and Tenements, unless he be able in true forme, measure, quantitie, and proportion, to plat the same in their particulars *ad infinitum*.

Certainly Agas practised what he preached. Among his own maps are one of Oxford with pictorial detail that includes the earliest known view of Oxford Castle, and one of the village and fields of Toddington in Bedfordshire that is among the finest of Elizabethan estate plans; we see here a small section only, for the entire map, on twenty sheets of parchment, measures some $11\frac{1}{2}$ by $8\frac{1}{2}$ feet (3·5 by 2·5 metres).

But this did not mean that by the end of the sixteenth century the role of the artist in map-making had come to an end. We have seen how the court painters in Franconia were continuing to produce picture-maps of estates in the early seventeenth century. And often enough surveyor and painter were combined in one person. August Hirschvogel, who, as we have seen, mapped Vienna by triangulation in 1547, was himself an artist. The surviving output of Jan

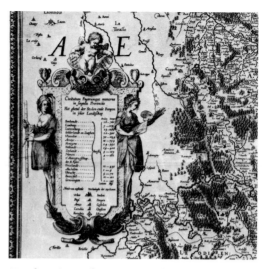

97 The science of measuring and the art of painting: allegorical figures on Willem Blaeu's map of the Seventeen Provinces of the Netherlands, 1608. The cartouche between them gives a key and lists the number of places in each province.

Potter, a Netherlands surveyor, includes not only five books of plans and forty-nine individual maps, dating from 1561 to 1590, but also two books of coats of arms that he produced in 1560. In 1631 Franz Kessler, a portrait painter of Cologne, certified that a large wall-map of Lindau that he had made had been accurately surveyed and painted. It comes as no surprise when we find that Arnold, the eldest son of Gerardus Mercator, not only, when young, helped his father in his surveys and subsequently succeeded him as director of his cartographic workshop, but was also an accomplished artist, who himself drew a bird's-eye view of Cologne for Braun and Hogenberg's *Civitates orbis terrarum*. For at the end of the sixteenth century the worlds of the map-making surveyor and of the topographical artist were still one. This was acknowledged by Willem Blaeu on the map of the Seventeen Provinces of the Netherlands that he published in 1608: a cartouche is flanked by two women, one representing the science of measuring, the other the art of painting. It neatly epitomizes the craft 97 of topographical mapping at that time.

11

The pictorial inheritance

WE SEE the two figures on Willem Blaeu's map of the Seventeen Provinces, one representing the science of measuring, the other the art of painting, as symbolizing topographical mapping in the early seventeenth century: traditions of picture-maps and of scale-maps based on survey existed side by side, and the practitioners of every method of representing landscape, cartographic or artistic, were often closely linked or even united in the same persons. But this is probably not quite what Blaeu had in mind when he had these two figures engraved on his map. Rather he will have meant to show that his own map was a work of collaboration between surveyor and artist, that the artist set out and embellished the map that had been drafted by the surveyor – the sort of collaboration that we have already seen in the map of the Vermandois that was designed by Van den Wijngaerde from the survey of Surhon in 1556. In such a partnership the surveyor had the major role: the information on the map, its outline and its detail, was his, and the artist had only to present the surveyor's work to its public in a form that was intelligible, perhaps also aesthetically beautiful. To Blaeu, as to us, a map was essentially a scale-map based on survey, the work primarily of the surveyor.

We can say in fact that it was in the early seventeenth century that the map and the bird's-eye view parted company. The picture-map of the sixteenth century and earlier lies behind two separate traditions of landscape representation, the cartographic tradition of the topographical map drawn to uniform scale, and the artistic tradition of the bird's-eye view drawn by the rules of perspective. When, in the introduction, we saw how the large-scale map of the area around Canterbury differs from an air photograph of the same area we were actually defining the difference between these two traditions; although in the air photograph the camera lens has taken the place of the artist's eye it presents us, obviously enough, with a bird's-eye view. As an artistic technique the view of landscape from above ground-level has had a distinguished history since the sixteenth century. We saw that detailed city views began in late-fifteenth-century Italy with the works of Rosselli and De' Barbari, and that the sixteenth century produced many others of this sort, culminating in the six-volume collection of *Civitates orbis terrarum* by Braun and Hogenberg. This tradition of bird's-eye views of towns continued, producing such masterpieces of intricately detailed work as the views of Rome by Giambattista Falda in 1676 or of Paris by Michel Etienne Turgot in 1739. Representing it here is a much later, much more modest production, but one of great charm and perhaps more typical of the tradition as a whole: a view by Thaddeus M. Fowler of El Reno in Oklahoma 98 in 1891, two years after its first settlement and a year before its first city charter, while it was still just starting the growth that was to give it a population of 3000 by 1900. Presenting an attractive picture that must have been meant to draw settlers to the new town it shows how for some purposes the bird's-eye view has advantages over the map. Another common use of bird's-eye views was for pictures of individual buildings. The seventeenth and eighteenth centuries produced a great many paintings or engravings showing bird's-eye views of country mansions; to the same tradition belong the sets of views of Oxford and Cambridge colleges that David Loggan drew and engraved in the 1670s and 1680s. Another very common use of the technique was for pictures of military operations; this again we can trace back to the

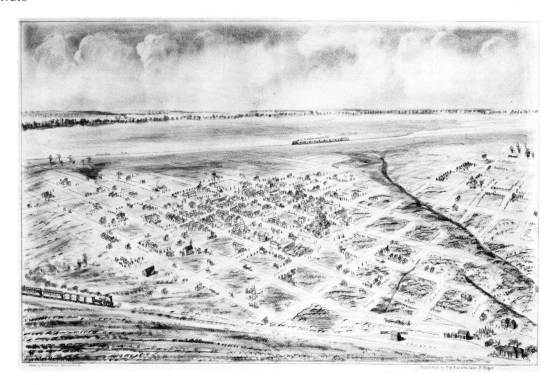

98 Bird's-eye view of a new town: El Reno, Oklahoma, in a lithograph by Thaddeus M. Fowler, 1891. Some blocks were already built up, others awaited building, and we see clearly the two railways serving the town.

56 beginning of the sixteenth century in the view of Lake Constance with scenes from the war of 1499, or even to the mid-fifteenth century in the incident from the war of 1437–41 shown on the district map of Verona. There are many other sixteenth-century examples, among them the magnificent set of engravings by Robert Adams to show the progress and defeat of the Spanish Armada in 1588 or, at a humbler level, 99 the view of the siege of Enniskillen in Ireland in 1594, one of a pair drawn 'by John Thomas, solder' (the other shows the battle of Ballyshannon in 1593). Large mural bird's-eye views of battles and sieges were often painted as decorations in royal palaces and public buildings in the seventeenth and eighteenth centuries.

At what date we start to distinguish between the two traditions, to say that this is a map, that

is a bird's-eye view, is more or less arbitrary. We have already seen that in the early seventeenth century there was among the court painters of Franconia a flourishing tradition of making estate maps in picture-map form, and there is no difficulty in matching their products from every part of Europe at the same period. Indeed there is no difficulty in tracing a tradition of picture-maps, combining artistic and cartographic elements in varying proportions, that has continued from that day to this. Publicity leaflets for tourists are a rich modern source of picture-maps on which we can find every characteristic of the picture-maps of medieval Europe and other societies. In a sense it is easy enough to separate our distinct traditions of maps and bird's-eye views: it is a map when it is drawn to a uniform scale, a view when it is drawn in strict perspective. But within these traditions we find relics of the common ancestry in the picture-map. In the bird's-eye view we can sometimes see this in the overall design where we find, not uncommonly, a departure from simple perspective in the use of double or multiple horizons. Occasionally we can see it in points of detail: just as we find a

100 miniature compass drawn as a direction pointer on Sengelaube's picture-map of the Truppach area in 1607, so we find exactly the same device on some of the undoubted bird's-eye views of parts of London that were engraved by Sutton Nicholls and published in *Several Prospects of the Most Noted Publick Buildings, in . . . London* (1724) and other collections. And, turning to the other tradition, we shall spend most of the rest of this chapter looking at features of topographical maps since the sixteenth century that stem from the picture-map.

But before we do this we must digress for a moment to look at a problem of definition. We have said that it is more or less arbitrary to choose a particular date for the separation of the two traditions of maps and bird's-eye views. But, more fundamentally, is it not entirely arbitrary to class Rosselli's view of Florence or De' Barbari's view of Venice as a picture-map, while speaking of Falda's representation of

Rome or Fowler's of El Reno simply as a view? Of course it is. In fact, rather than seeing our medieval and Renaissance picture-maps as giving rise to a double tradition of maps on the one hand, bird's-eye views on the other, it might well be more realistic, more in keeping with the ideas of the age that produced them, to see them simply as the beginnings of an artistic tradition of landscapes viewed from above ground-level, a tradition which culminated in the detailed and accurate bird's-eye views of the Renaissance and later but which also produced, as an odd offshoot, the topographical map

99 A vivid, if rather confused, picture-map of the siege of Enniskillen, 1594. Nothing is known of the artist, who signs himself as John Thomas, soldier. Maps of military actions were common in the late 16th century and often, as here, they were a cross between a picture-map and a picture-story.

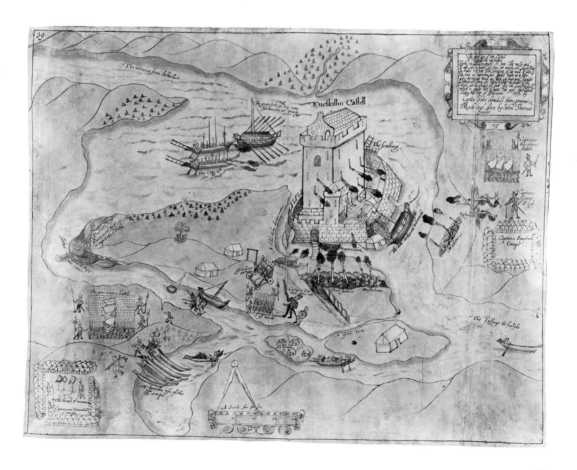

100 Picture-map of the Truppach area, near Bay-reuth, made in 1607 by Peter Sengelaube, architect and court painter of Saxe-Coburg, for a lawsuit over hunting rights. The picture of a compass, with south-pointing needle, is a direction pointer.

which replaced perspective by the cartographic convention of uniform scale.

Just what, after all, is a map? In the introduction we distinguished the map from the air photograph (or bird's-eye view) by simply this: uniformity of scale throughout, and therewith vertical representation at every point. Then, when we turned to look at the most primitive symbol-maps, we defined as a map any design or group of objects which represented features of landscape and which in its arrangement reflected the arrangement of those features on the ground. This is an extremely wide definition. It would bring within its scope all bird's-eye views (and air photographs), and a great deal else besides: all picture-maps and diagram-maps, all pictures of landscape even if viewed from ground-level, and any written list of places along a route, for, after all, what could better represent a feature of landscape than its own name in writing? Obviously this definition, appropriate enough for the symbol-maps of primitive peoples, is far too wide to be used in other contexts, and in fact – as the reader may or may not have noticed – we have silently adjusted our idea of what is and what is not a map as we have moved to different cultures and different ages. In fifteenth-century England we looked at Robert Cole's survey of Gloucester, with its parallel lists of house-owners along the streets and its thumb-nail sketches of other features, and decided that it is not quite a map – a ruling in its way no less arbitrary than the exclusion of Falda's view of Rome or Fowler's view of El Reno.

What then is a map? The answer is a very simple one. It is exactly what we found it to be when we contrasted map and air photograph; until the emergence of the surveyed map, drawn (or intended to be drawn) to a uniform scale, there was no such thing as a topographical map. What we have done in this book is to call maps those representations of landscape that seem to lie in a line of development (typological, not necessarily historical) leading to the real topographical map. This has been a convenient shorthand, but it may be misleading. When we speak of symbol-maps, diagram-maps, picture-maps, we are not really speaking of maps at all – they are simply the nearest approach that particular societies achieved to the concept of cartography that was attained only with the production of the scale-map. We can see each of them as a sort of map only because we are looking at them with hindsight, because we know what a real map is. The men who made them knew nothing of maps; to them they were simply topographical symbols, topographical diagrams, a particular sort of topographical picture. It is significant that there was no word for a map in any European language until the Renaissance; there was no word for a map because maps did not exist. This does not mean for a moment that we are wrong to make use of our hindsight, that we should not try to distinguish these particular symbols, diagrams, pictures, that seem to lead the way to the map; on the contrary, the way mankind reached an understanding of cartography, how it works and how it can be used, is of extraordinary interest in the history of both artistic and scientific ideas. But we must recognize that many of our classifications, many of the lines of development that we have distinguished, are entirely artificial in the context of their own time; they are significant only in relation to the surveyed scale-map, the real map that made its appearance in Babylonia, in the Roman Empire, in Chin dynasty China and finally in Renaissance Europe.

But even when the real topographical map, based on survey and drawn to scale, was fully established in the tradition that began in sixteenth-century Europe, there was still much about it that reminds us of its pictorial ancestry. Let us look at the map of Philadelphia by Charles Varlé that was published in 1802. It was, it must

101

be admitted, a rather old-fashioned map for the early nineteenth century, but this makes it all the more suitable for our purpose. Basically it is a straightforward scale-map of the city that was the second largest in the United States and had just ceased, two years before, to be the national capital; but added to this basic plan are various pictures that contribute a great deal to the map's visual interest and quaint charm. These pictorial elements in the map are of three sorts. First there are those that are pure embellishment, decorative pictures that contribute no additional information to the plan itself: the ships on the Delaware River, the pictures of the city's principal buildings along the map's bottom edge. Then in parts of the map the simple plan is supplemented by a view of the landscape; the area outside the city to the north is shown not as a map but as a picture, much conventionalized, of its houses, fields and plantations. Finally some of the pictures on the map can be seen as conventional signs pure and simple; the trees in the town and outskirts fall in this category. These three pictorial elements in the map – the picture as decoration, the picture as an integral part of the map, the picture as conventional sign – represent the three ways a pictorial tradition has continued in topographical mapping from the sixteenth-century picture-map down to the present. The distinctions between the three may be even less clear-cut than on Varlé's map, especially when we look at their early development, but they form a useful broad classification of the subject. An entire book could be written on the development of any of the three and, indeed, there are already many books and articles on what are in effect aspects of the pictorial tradition in mapping. All we can do here, by way of rounding off our story, is to look at each of the three in turn so as to see what relics of the picture-map, what traits derived from a pictorial ancestry, are to be found in the topographical map of today.

Overleaf

101 Philadelphia in 1802. A map by Charles Varlé that exemplifies the three continuing pictorial traditions in mapping: the picture as embellishment, the picture as an integral part of the map and the picture as conventional symbol.

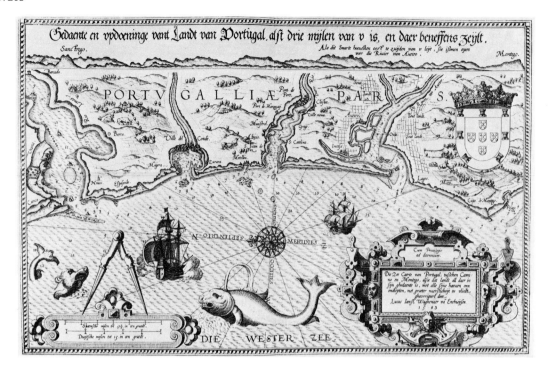

Gedaente en vpdoeninge van Landt van Portugal, alst drie mijlen van v is, en daer beneffens Zeijlt.

PORTVGALLIÆ PARS.

DIE WESTER ZEE.

102 The coast of north Portugal in the sea-atlas of Lucas Waghenaer, 1583. Despite its decorative qualities this is a practical map: soundings are given near the shore and up the estuaries, and along the top is a coastal profile.

It is not in the least difficult to find sixteenth-century maps that include pictures by way of decoration. The area of a map might be covered with imaginary views of landscape, like Eufrosino della Volpaia's engraved map of the Roman Campagna of 1547 or the mural maps of the papal territories that were painted in the Vatican Palace in the 1570s and 1580s. What we see here is more restrained, but at the same time it is a sufficiently striking example of this sort of embellishment, striking partly in the simple charm of the pastoral landscape it envisages in north-west Portugal, partly in appearing on a severely practical map: it is one of the great collection of coastal charts in the earliest published sea atlas, Lucas Waghenaer's *De spieghel der zeevaerdt* (1583; the title became *The Mariner's Mirror* in the English version of 1588). In the sixteenth century the fact that a map was to serve a practical use was no reason for

failing to embellish it and, indeed, down to the early eighteenth century a map would normally include landscape views and pictures of animals and people typical of the regions shown. Historians of cartography are apt to quote from Jonathan Swift's poem 'On poetry' (1733):

> So Geographers, in *Afric*-Maps,
> With Savage-Pictures fill their Gaps;
> And o'er unhabitable Downs
> Place Elephants for want of Towns.

But it was water – sea, rivers and lakes – that offered the most embarrassing blank spaces, and it was here that pictorial decorations persisted longest in formal cartography, though with growing restraint: the ships that we see on Varlé's map of Philadelphia or on the 'Duke's map' of New York in 1664 are picturesque enough, but they are a tame substitute for the great fish on Waghenaer's map, let alone the figure of Poseidon on De' Barbari's view of Venice.

This same tendency to restraint and formality led to scenic embellishments being moved from the surface of the map to the borders; here eighteenth-century maps often gave imaginary views of landscape as decorative vignettes. Sometimes, interestingly, these included pictures of the surveyor at work. From England we see an early instance on some of John Ogilby's road maps of 1675; others are on printed county maps, such as Henry Beighton's Warwickshire in 1725, or on manuscript estate maps, as on a draft vignette for John Rocque's survey in 1757 of the Earl of Kildare's lands (in the final version the surveyor and his assistant were replaced by a horse and cart). Again, in the course of the eighteenth century this same move away from sheer exuberance of decoration produced views of actual landscapes and monuments instead of imaginary scenes in the borders of maps. The first series of English county maps to do this was drawn up in the 1750s for *The Large English Atlas* by Emmanuel Bowles and Thomas Kitchen, and during the next hundred years many others followed suit – the two series in Pigot and Co's directories in the 1820s and Thomas Moule's *The English*

105

Counties Delineated, first published in 1836, have notably fine engraved vignettes of local buildings and views. We find the same fashion on estate maps, which might well include in the borders sketches of the owner's mansion or other views of the property. But by the mid-nineteenth century even this form of pictorial decoration had disappeared from formal cartography: there is no trace of it on the modern official maps from Britain and Switzerland that we looked at in the introduction, nor yet on their nineteenth-century predecessors. It does of course still persist – as picture-maps themselves persist – on the less formal mapping of travel posters, tourist literature and the like. But

103 The 'Duke's map' of New York, 1664. Detail is shown pictorially on a ground-plan true to scale; we see clearly the Battery and the town wall that Wall Street commemorates.

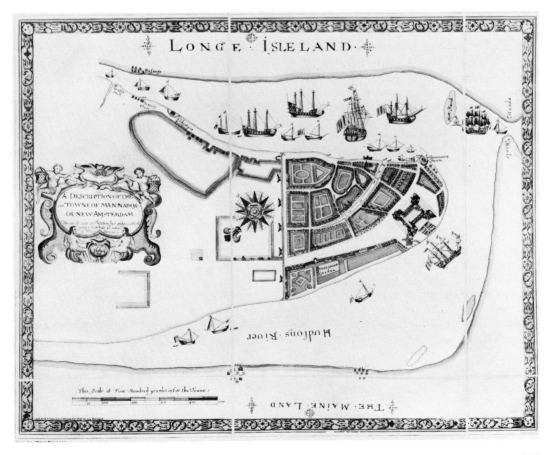

177

though the continuance of this pictorial decoration is an interesting link with the picture-maps of the past it is less significant than the two other pictorial elements that we found on Varlé's map of Philadelphia, for it is mere embellishment, not truly part of the map itself.

When we turn to the second of these pictorial elements we are looking at something much more fundamental, for this is the picture used as an integral part of the map: detailed features are shown not by conventional signs nor by their outline ground-plans, but by pictures of their actual appearance as viewed from above at an oblique, near vertical, angle. At first sight this seems to be simply the technique of the bird's-eye view, and we might say that Varlé's plan of Philadelphia is a mixture of bird's-eye view and map. This would be incorrect. The features that appear pictorially are not drawn to a single overall perspective; there is no general foreshortening with increasing distance, no single notional horizon. It is not a bird's-eye view. Rather it is a multiplicity of bird's-eye views, a collection of separate pictures of individual features and complexes placed in position on a ground-plan that remains true to scale throughout. Each feature has its own perspective, its own horizon; but because the angle of view is not far from vertical, the divergence from real bird's-eye view with a single very distant horizon is quite small, just as the near-vertical air photograph produces a pattern of landscape that differs only minutely from that of a map. We see this technique particularly clearly on a map of London by John Overton that was produced immediately after the Great Fire of 1666; the areas destroyed in the fire are shown simply as blank blocks, while the areas untouched have houses and other buildings drawn in. But we see that it is a single plan, drawn to uniform scale, on which the pictures of buildings have merely been superimposed; there is no overall foreshortening on the pictorial portions of the map. On the other hand because the scale is small and the angle of view not far from vertical one needs more than a mere glance

104

104 Map of London drawn by Wenceslaus Hollar and published by John Overton immediately after the Great Fire of 1666. Areas destroyed are shown by blocks left blank, those untouched by pictures of buildings drawn in.

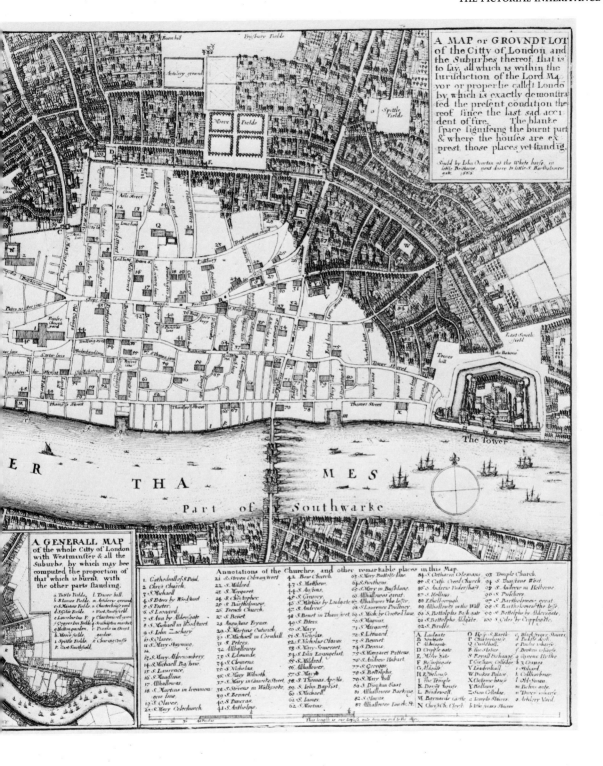

A MAP or GROVNDPLOT of the Citty of London and the Suburbes thereof, that is to say, all which is within the Iurisdiction of the Lord Mayor or properlie called London by which is exactly demonstrated the present condition thereof since the last sad accident of fire. The blanke space signifeing the burnt part & where the houses are exprest, those places yet standig.

Sould by Iohn Overton at the White house in little Britaine next dore to little S. Bartholomew gate. 1666.

A GENERALL MAP of the whole Citty of London with Westminster & all the Suburbs, by which may bee computed the proportion of that which is burnt, with the other parts standing.

at the map to discover that it is not drawn as a genuine bird's-eye view to a single overall perspective.

The roots of this technique are bedded deep in the history of the picture-map. It could be argued, for instance, that this type of representation appears in a fairly developed form in the plan of Verona on the mid-fifteenth-century district map. It is, after all, no more than a sophisticated extension of the multiple perspective which is a technique common on picture-maps of every tradition, not only those of medieval Europe, and which we can find on scale-maps at the beginning of the sixteenth century; we have met it already on the Black Forest map by Tibianus. Detailed views of towns in the sixteenth century were drawn sometimes on this method, sometimes as true bird's-eye views; Jörg Seld's view of Augsburg in 1521 is a bird's eye view, like De' Barbari's Venice in 1500, whereas the view of Antwerp by Virgilius van Bologne and Cornelius Grapheus in 1565 is really a multiplicity of pictures superimposed on a true scale-map. The technique could be easier than that of the bird's-eye view, as it eliminated the overall foreshortening in which even De' Barbari had not achieved complete accuracy.

But apart from these monuments of detailed pictorial mapping, the same technique occurs on an enormous number of topographical maps of every sort in the sixteenth, seventeenth and eighteenth centuries. Sometimes it is used just for a few features, such as the fences, hedges and gates on an estate plan, sometimes it is applied systematically to all the detail of a map. In the illustration we see it on the 'Duke's map' of New York, 'A description of the towne of Mannados or New Amsterdam as it was in September 1661', a map drawn in 1664 from drafts made on the town's capture from the Dutch three years before; this is a scale-map of the south end of Manhattan Island, showing the town's wall, fortress and streets, but superimposed on it are perspective views of the rows of houses and other buildings. Sometimes the technique is used with detailed precision, sometimes casually with no particular regard for either pictorial accuracy or for scale, so that the picture can become little more than a conventional sign. This may be the case with the windmill on the New York map. It is certainly the way pictorial representation was used on the road maps of

England and Wales that John Ogilby published in his *Britannia* in 1675. This was Britain's first road atlas, and it also introduced the scale of one inch to one mile (1:63360) that became so common among British map-makers. In the introduction to *Britannia* Ogilby explained the system he used:

Capital Towns are describ'd *Ichnographically*, according to their Form and Extent; but the *Lesser Towns* and *Villages*, with the *Mansion Houses, Castles, Churches, Mills, Beacons, Woods*, &c. *Scenographically*, or in *Prospect*.

We see one of Ogilby's maps in the illustration, showing the road from London to Oxford and Islip, the first part of the road to Aberystwyth. Oxford and London are shown by straightforward plans, Uxbridge, High Wycombe and other places by tiny rows of houses lining their main streets; and other features of the roadside landscape appear as pictures, conventional in form. The system was well adapted for this particular purpose, and it is interesting that it is the exact opposite of many fifteenth- and sixteenth-century maps which show only towns pictorially.

But any number of examples could be produced of a technique that was part of the normal stock-in-trade of a maker of large-scale maps anywhere in Europe before the nineteenth century. An interesting epilogue has recently been brought to light by Mr Ralph Hyde. In 1813 Thomas Hornor, a young English surveyor, having produced specimens of a method of map-making that he called 'panoramic chorometry', published a book entitled *Description of an Improved Method of Delineating Estates*. Here he explained how

The arts of surveying and landscape painting, which seem to have been united in former days, are now distinct. That a plan may be drawn with the same precision and afterwards so finished as to form a faithful and interesting picture of the various features of the property . . . has been proved by the enlarged specimens which I recently admitted to public inspection. In these the whole subject country is represented in the colours of nature, and all its parts are drawn in a correct and faithful manner.

And in the following years he put his method into practice in a series of plans of estates in the Vale of Neath. It consisted of pictorial

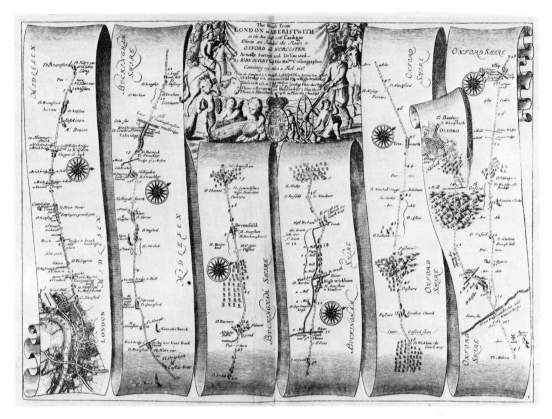

105 The road from London to Oxford and Islip, the first part of the route to Aberystwyth: a road map from John Ogilby's *Britannia*, 1675. These maps show landmarks along the roads, but no other features outside towns; the compass roses mark the changes of direction from one strip to the next.

mapping taken to its extreme limit of thoroughness: on his scale-map of the area he superimposed a near-vertical picture of the landscape, as detailed as an air photograph would be but, of course, differing from the air photograph in its total uniformity of scale, its lack of a single overall perspective, and its maintenance of a constant angle of view at every point. The result was a representation of landscape that was faithful and effective and even – for Hornor was an accomplished artist – beautiful; but the time spent in this artistic work must have added considerably to the expense of a survey and in any case the tide was running fast against Hornor. In his book he remarks that a pictorial approach to map-making was out of favour with contemporary surveyors, and indeed by the

early nineteenth century pictures on formal maps had been generally abandoned for the austere simplicity of outline ground-plan and conventional symbols.

But with one exception: the depiction of relief. This was probably the hardest problem that map-makers had to solve in advancing from the picture-map to the scale-map. To the artist of the picture-map there was no serious difficulty about the way hills and mountains were to be shown; he nearly always drew them as seen from the side, either from ground-level or, like other features on the map, from a moderate height. The actual shape of the mountains on the map naturally varied according to the particular artistic style, from the multiple inverted *V*s of the Chinese woodcuts to the signs like crouching snouted animals on Mexican maps and the realistic pictures of peaks and valleys on some of the fifteenth-century maps from north Italy. The method had the disadvantage of masking what lay immediately behind the mountain range if the pictorial conventions were properly observed. Leonardo da Vinci seems to have been

181

concerned with the problem of how to show on his maps rivers that were hidden in valleys, and at a cruder level of picture-map we have seen a solution attempted by opening up a landscape along a valley to show both sides in facing views with opposite horizons. More problems arose when the same method of showing hills in elevation was transferred to scale-maps. When, as for instance on Henry Beighton's large-scale map of Warwickshire in 1725, we find a crossroads on a hill represented by superimposing the road pattern on a picture of the hill seen from the side, the result is decidedly odd. John Ogilby on his road strip-maps of 1675 shows where roads go uphill by normal pictures of hills, and where they go downhill by pictures of hills upside down – a clumsy device that reaches absurdity where (as we see in two instances here) the reversed hills are covered with pictures of trees representing woodland but drawn the right way up.

In England this method of showing relief produced minor inelegancies but no serious problems. It was otherwise in countries where the landscape was dominated by high mountains. Some small-scale maps of the sixteenth century show mountains simply by outlining and shading their ground-plan, but this was not satisfactory, for it was impossible to show details of the mountains themselves or of any other features within those areas, and in the mid-eighteenth century straightforward pictures, more or less conventionalized, either in elevation or in oblique perspective, were still the normal way of showing mountains on maps. The device of showing the farther side of mountains upside down was used even for the Alps, as in 1781 in a book by M. T. Bourrit, *Description des Alpes Pennines et Rhétiennes.* Even the large-scale map of France by the Cassini family, completed in 1793, in many ways the most advanced cartographic production of its time, still used simple oblique pictures to mark peaks, shoulders and large outcrops of rock; the way it shows mountain areas was described in 1907 by H. Béraldi:

Among broad, flat valleys are broad, flat plateaux on which, to mark outcrops, are placed little pieces of stone like *croûtons* on a *purée*; here and there, marking exceptional peaks . . . a piece of *pâtisserie*, a *moule à bavarois.*

The description underrates the merits of map-makers who were among Europe's greatest, but it underlines the difficulties and shortcomings of their way of showing relief.

One solution in keeping with the general cartographic trends of the time was to use contour lines to mark relief, abandoning pictorial representation of hills altogether. This was the application to dry land of a system developed by marine cartographers from the 1730s on to show depth of water; it was put before the French Academy in 1771 as a way to show relief, but it was not until twenty years later that it was used on a map of France and only in the second quarter of the nineteenth century that it came to be widely adopted. It may seem strange that the use of contour lines spread so slowly, for at first sight they have obvious advantages over any pictorial method of showing relief: they are clear and precise and they keep the surface of the map relatively uncluttered. On the other hand they give a much less striking impression of relief than pictorial methods – in non-mountainous country one has to read the figures on successive contours to discover the direction of slope – and in mountain areas where there are high peaks and vertical rock-faces the contour lines merge or become so closely spaced that the system more or less breaks down. Here there are clear advantages in a sophisticated pictorial method.

Such a method had already been devised and elaborated before contour lines were used on land maps. This was hachuring: lines or shading to show the exact shape of mountains and hills as seen from a vertical viewpoint. It has been said that the map of Breisgau published by J. B. Homann of Nuremberg in 1718 was the first to use systematic hachuring, but it can be found in rudimentary form on earlier maps. The earliest hachuring consisted simply of lines drawn from the top of the peak or ridge following the path that would be taken downwards by freely flowing water, but in the mid-eighteenth century various more elaborate systems were in use. Deeper shading, produced by drawing the lines broader or closer together, could be used to show either height or degree of slope – either the higher the darker or the steeper the darker – or to make the relief stand out more sharply by marking the areas that would be in shadow when the sun was at a particular point. In 1799 a

105

book by J. G. Lehmann, *Darstellung einer neuer Theorie der Bezeichnung der schiefen Flachen* ('Exposition of a new theory of marking relief'), was published in Leipzig. This was a turning-point; while it drew on the work of earlier cartographers, it succeeded in elevating hachuring to the status of a precise system that was both complex and subtle. But its fundamental basis remained pictorial; thus, assuming vertical illumination, Lehmann argued that the light rays would be less concentrated where the slope was steeper, and so used the deepest shading to mark the steepest slopes. Methods of hachuring have continued to develop since Lehmann's time. Horizontal instead of vertical hachure lines first appeared in 1826 on an official map of Norway, and attempts have been made to get the best of both systems by drawing contour lines heavily or lightly in the manner of hachuring. During the last century there have been many maps that use contour lines and hachuring together to show relief – among them the official 1:50000 map of Switzerland that we looked at in the introduction. Here contour lines give us the precise heights and shapes of the lower slopes and glaciers but the highest peaks and ridges are marked by hachuring to give us a clear visual image of their form. This use of hachuring is the last relic in formal topographical mapping of the picture as an integral part of the map and it links our modern Swiss large-scale map with the picture-map traditions of the sixteenth century and earlier periods.

II

The third pictorial element that we found on Charles Varlé's map of Philadelphia was the picture used as conventional sign. This again goes far back into the history of the picture-map. We have seen how the city ideogram in its various forms links the picture-map of classical antiquity with that of the late middle ages. In fact we have mostly passed over in silence how far the detail on picture-maps is realistic, how far conventionalized; this offers a vast field of investigation to the historian of art as well as of cartography. Given that much of the art of medieval Europe was symbolic rather than realistic it is not surprising that its picture-maps should often be strongly conventionalized. But the conventions seem always to have had a pictorial basis, and the only abstract symbol that occasionally occurs on medieval topographical

maps is the dot or tiny circle to mark towns. Interestingly, perhaps significantly, conventional signs that are not pictorial start to multiply only with the spread of scale-maps in the sixteenth century. We have seen how in 1578 the map of the Black Forest by Tibianus marked abbeys by a crozier. On the map of Bohemia by Nicholas Claudianus that was published in 1518 we have a more elaborate use of symbols: Hussite towns were marked with a chalice and Catholic ones with crossed keys, while royal possessions were distinguished from feudal by crowns and horses' heads. Once symbols stopped being simply pictorial, some explanation of their meaning was needed, and the earliest map with a key was one of Franconia in 1533 by Peter Apianus. But among conventional signs the abstract and the pictorial cannot always be sharply distinguished. John Ogilby in his map of Kent marked towns where markets or fairs were held by a cross and circle at the end of a staff, a sign that may represent the bush hung outside taverns; the familiar crossed-swords sign to mark the site of a battle derives from actual pictures of soldiers fighting.

1

X

But if it is difficult to tell the abstract sign from the pictorial one it is often far harder to distinguish conventional sign from would-be pictorial representation. Even in so late a map as Varlé's plan of Philadelphia our distinction between the two is open to question, and we have seen how John Ogilby's method of showing landscape 'scenographically' on his road maps was really no more than a system of conventional signs. Conversely we sometimes find that where we expect conventional signs we have in fact real pictures. In the 1490s Konrad Türst's map of Switzerland marked towns and villages by thumb-nail sketches of their actual appearance, and from his time to ours some maps have used actual pictures to mark features. Often this is obvious: a map of Bohemia in 1737, for instance, shows the sites of Franciscan friaries by carefully engraved pictures. Sometimes it is not obvious at all. In the illustration we see one of the maps of the four Warwickshire hundreds that were drawn by William Dugdale and published in 1656 in his history of the county; the parish churches are marked by tiny pictures which are not conventional signs but show their actual appearance at the time.

106

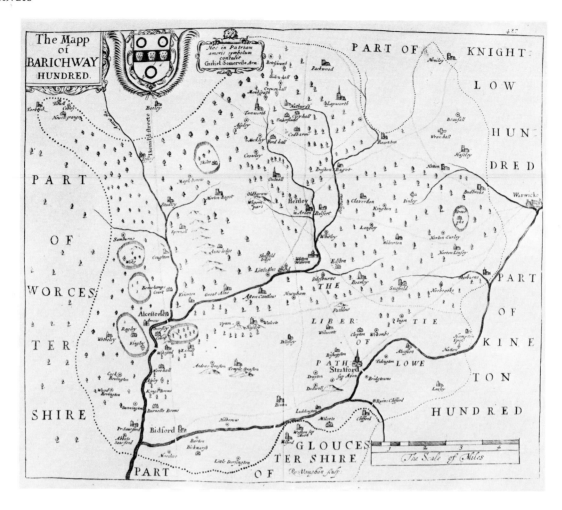

106 Barlichway ('Barichway') Hundred, south-west Warwickshire, drawn by William Dugdale for his *Antiquities of Warwickshire*, 1656. The buildings marking towns and parishes were drawn from life: Warwick Castle (centre right) and Stratford-upon-Avon church (bottom centre) can be clearly recognized.

In conventional signs the closest link of all between the modern topographical map and its picture-map ancestors – a link so obvious that it can easily be overlooked – is the use of colour. Distinguishing arable land from pasture by brown and green, colouring water blue, roads brown or yellow, buildings and towns grey, pink or red, all these are conventions so persistent as to have been almost universal in the topographical maps of the last five hundred years. If in the introduction we had compared our modern map of the area around Canterbury with an air photograph not in black and white but in full colour we would of course have found great differences in the colour-values of particular features and areas, but equally we would have found a significant relationship in the overall colour scheme. It is in using conventional signs of pictorial origin that the modern British Ordnance Survey map is most closely linked to the picture-map of the past, and in nothing is this clearer than in its use of colour.

So, having returned to our modern maps of the area around Zermatt and the area around Canterbury we are back at our starting-point.

We have ranged widely in space, widely in time, but much of the story of how man learned to make maps, how to see the landscape cartographically, must have occurred in ages and in regions from which no record survives. In 1897 Frederick William Maitland, the great historian of the middle ages, with characteristic insight described the British Ordnance Survey map as a 'marvellous palimpsest' and remarked that 'Two little fragments of "the original one inch ordnance survey map" will be more eloquent than would be many paragraphs of written discourse'. He meant that the topographical map in its record of the modern landscape is, if properly understood, a record also of successive stages of the historical process that brought that landscape into being. His advice has been heeded, and our knowledge of the society and economy of past ages has been greatly broadened and deepened as a result. But his comment could apply equally well to the history of cartography itself; our map contains many relics of successive stages in that history and read aright they can help us to an understanding of how its ideas and techniques came into being. Here we have taken a few first steps in that direction; much more still remains to be explored.

Notes on authorities

Introduction

Of general histories of cartography, the best available is L. Bagrow, *History of Cartography*, ed. R. A. Skelton (1964); it deals particularly well with the developments in Europe in the fifteenth to seventeenth centuries. A useful brief survey is provided by the two chapters, by C. Singer, D. J. Price and E. G. R. Taylor, on 'Cartography, Survey, and Navigation' before 1400 and from 1400 to 1750 in *A History of Technology*, ed. C. Singer and others (7v., 1954–78), iii, pp. 501–57. R. V. Tooley and C. Bricker, *A History of Cartography* (1969), is especially noteworthy for its many fine illustrations of early maps.

The five quotations on p. 12 are taken from, respectively, A. Cortesão, *History of Portuguese Cartography* (2v., 1969–71, i, p. 82; E. Oberhummer, 'Der Stadtplan, seine Entwickelung und geographische Bedeutung', *Verhandlungen des sechszehnten Deutschen Geographentages zu Nürnberg* (1907), p. 71; J. Lelewel, *Géographie du moyen âge* (4v. and epilogue, 1852–7), i, p. 91; F. Geerz, *Geschichte der geographischen Vermessungen und der Landkarten Nordalbingiens vom Ende des 15. Jahrhunderts bis zum Jahre 1859* (1859), p. 38; Tooley and Bricker, op.cit., p. 85. I cannot emphasize too strongly that whereas I disagree with the assumptions underlying these particular quotations I have great respect for the knowledge and scholarship of all these authors and have indeed made substantial use of their work in the present book.

Ptolemy draws his distinction between geographic and chorographic maps in his *Geographia*, book I, chapter 1 (ed. C. F. A. Nobbe (3v., 1843–5), i, pp. 3–5). The eighteenth-century household inventory quoted is from the Probate Records of the diocese of Llandaff; I owe it to the kindness of Dr J. B. Harley. The quotation on cartographic optics is from K. Peucker, 'Zum kartographischen Darstellung der dritten Dimension', *Geographische Zeitschrift*, vii (1901), p. 34.

The work by D. Wood is published as 'Now and then', *Prologue: the Journal of the National Archives*, ix (1977), pp. 151–61.

1   The beginnings

The few general accounts of the maps of primitive peoples, although they are far from recent and make little attempt at analytical discussion, are valuable in bringing together many scattered references to maps from the works of early explorers and other travellers: R. Andree, 'Die Anfänge der Kartographie', *Globus*, xxxi (1877), pp. 24–7, 37–43; W. Dröber, *Kartographie bei den Naturvölkern* (1903), of which a summary, under the same title, appeared in *Deutsche Geographische Blätter*, xxvii (1904), pp. 29–46; H. de Hutorowicz, 'Maps of primitive peoples', *Bulletin of the American Geographical Society*, xliii (1911), pp. 669–79, itself a summary of the work in Russian by B. F. Adler, *Karty piervobytnyh narodov* (1910). Much of the information used for this chapter and the next is drawn from one or other of these works.

The spear-throwers of the Bindibu are described by D. F. Thomson, 'The Bindibu Expedition', *Geographical Journal*, cxxviii (1962), p. 274. The fullest account of the Marshall Islands stickcharts is A. Schück, *Die Stabkarten der Marshall-Insulaner* (1902), but W. Davenport, 'Marshall Islands Navigational Charts', *Imago Mundi*, xv (1960), pp. 19–26, gives a newer interpretation; the work by Schück includes (pp. 34–6, pl. 8–10) A. C. Haddon's account of the stones on Mer. K. von den Steinen's descriptions of the maps drawn for him in Brazil are in his *Durch Central-Brasilien* (1886), pp. 213–14, and *Unter den Naturvölkern Zentral-Brasiliens* (1894), pp. 153, 246–7.

Descriptions of maps by North American Indians are included in the discussion of Indian picture-writing by H. R. Schoolcraft, *Historical and Statistical Information Respecting the . . . Indian Tribes of the United States* (6v., 1851–60), i, pp. 333–430, but C. A. Burland, 'American Indian Map Makers', *Geographical Magazine*, xx (1947), pp. 285–92, is a simpler introductory account. D. Norona, 'Maps Drawn by Indians in the Virginias', *The West Virginia Archeologist*, vii (1950), pp. 12–19, gives a full list of those recorded from one region; the

2   From symbol to picture

The general accounts of the maps of primitive peoples listed in the note dealing with chapter 1 have been drawn on for this chapter as well.

The carved maps of the Greenland Eskimos are described in 'The Ammassalik Eskimo: Contributions to the Ethnology of the East Greenland Natives', ed. W. Thalbitzer, in *Meddelser om Grønland*, xxxix (1914), pp. 107–8, 665–6. Wilford's description of the map of Nepal is from 'An Essay on the Sacred Isles in the West', *Asiatick Researches*, viii (1805), p. 271; the accounts of the Inca maps in relief are from P. Sarmiento de Gamboa, *History of the Incas*, ed. C. Markham (Hakluyt Soc., 1907), p. 120, and G. de la Vega, *First Part of the Royal Commentaries of the Yncas*, ed. C. R. Markham (Hakluyt Soc., 2v., 1869–71), i, p. 190; and the quotation from Ibn Battúta's work is from *Ibn Battúta: Travels in Asia and Africa 1325–1354*, ed. H. A. R. Gibb (1929), pp. 312–13. The map described showing the area around the River Oxus is reproduced to accompany J. Bird, 'An Account of the City of Balkh and its Neighbourhood, Extracted from Persian Authorities', *Transactions of the Bombay Geographical Society*, ii, part 1 (Aug. 1838), pp. 60–71 (the map is at the end of the part); a substantial collection of Arab maps of particular regions and wider areas, dating from the ninth century to the thirteenth, has been published as *Mappae Arabicae: arabische Welt- und Länderkarten*, ed. K. Miller (6v., 1926–31). The evidence for maps in relief from China is discussed by J. Needham, *Science and Civilisation in*

The quotation from Norwood is from A. and J. Churchill, *A Collection of Voyages and Travels* (6v., 1732), vi, p. 165, as cited by Norona, p. 15. But on the cartography of North American Indians we await the forthcoming definitive work of Dr G. M. Lewis.

R. J. Flaherty's account of the map of Wetalltok and his exploration of the Belcher Islands is 'The Belcher Islands of Hudson Bay: their Discovery and Exploration', *Geographical Review*, v (1918), pp. 433–58.

China (in progress, 1954–), iii, pp. 579–83; the quotation from Shen Kua is taken from p. 580, the description of the Ch'in dynasty tomb from p. 582.

Works on the maps of North American Indians are given in the note dealing with chapter 1; Schoolcraft's description of the Ojibwa birchbark message is from his *Historical and Statistical Information*, i, p. 336. For the Early Han dynasty maps from Chang-sha see the note dealing with chapter 6. The carvings of the Valcamonica are discussed, with many illustrations, by E. Anati, *Camonica Valley* (1964).

3 The classical tradition

The comment by E. G. R. Taylor at the beginning of the chapter is from a discussion on early maps of Britain, recorded in *Geographical Journal*, lxxxi (Jan.–June 1933), p. 44.

The clay-tablet map from Nuzi is described and discussed by T. J. Meek, *Old Akkadian, Sumerian and Cappadocian Texts from Nuzi* (1935), pp. xvii–xviii. An illustrated account of the map painted on coffins from ancient Egypt is given by W. Bonacker, 'The Egyptian "Book of the Two Ways"', *Imago Mundi*, vii (1950), pp. 5–17. A good deal has been written on the Turin papyrus map; the definitive work is by G. Goyon, 'Le papyrus de Turin dit "des mines d'or" et le Wadi Hammamat', *Annales du Service des Antiquités de l'Egypte*, xlix (1949), pp. 337–92. Egyptian plans of houses and gardens are reproduced in J. G. Wilkinson, *The Manners and Customs of the Ancient Egyptians* (3rd edn, 5v., 1847), ii, p. 5, opp. p. 94, p. 105, 129–48, and in J. Capart, *Lectures on Egyptian Art* (1928), pp. 86–91.

The painting from Thebes of an army drawn up before a town is reproduced in Wilkinson, op. cit., i, p. 382. The picture-map bas-reliefs from the Assyrian Empire are described in G. Perrot and C. Chipiez, *A History of Art in Chaldaea and Assyria*, ed. W. Armstrong (2v., 1884), i, pp. 329–33. The Greek coin maps from Zancle are described and illustrated by C. M. Craay, *Archaic and Classical Greek Coins* (1976), p. 207, pl. 44, those from Ephesus by A. E. M. Johnston, 'The Earliest Preserved Greek Map: a New Ionian Coin Type', *Journal of Hellenic Studies*, lxxxvii (1967), pp. 86–94, pl. IX–XI.

The best general account of picture-maps from the Roman Empire is in A. and M. Levi, *Itineraria picta: contributo allo studio della Tabula Peutingeriana* (Museo dell'Impero Romano, 1967), pp. 17–64. The maps in the Roman treatises on surveying are discussed in two articles by O. A. W. Dilke, 'Maps in the Treatises of Roman Land Surveyors', *Geographical Journal*, cxxvii (1961), pp. 417–26, and 'Illustrations from Roman Surveyors' Manuals', *Imago Mundi*, xxi (1967), pp. 9–29. The fullest account of the Madaba mosaic map is M. Avi-Yonah, *The Madaba Mosaic Map* (Israel Exploration Soc., 1954).

Medieval maps of the Holy Land are discussed and illustrated in a series of articles by R. Röhricht in *Zeitschrift des Deutschen Palästina-Vereins*: 'Karten und Pläne zur Palästinakunde aus dem 7. bis 16. Jahrhundert', in six parts (xiv (1891), pp. 8–11, 87–92, 137–41, pl. 1, 3–5; xv (1892), pp. 34–9, 185–8, pl. 1–9; xviii (1895), pp. 173–82, pl. 5–7), 'Die Palästinakarte des William Wey' (xxvii (1904), pp. 188–93); see also the articles on the maps in Breitenbach's and Sanudo's works that are cited in the notes dealing with chapters 4 and 9. The maps by Matthew Paris are also discussed by R. Vaughan, *Matthew Paris* (1958), pp. 241, 244–7, pl. XVI, XVII.

Several of the medieval district maps from north Italy as well as the fifteenth-century maps of Tuscany and the Naples boundary map are reproduced and discussed in R. Almagià, *Monumenta Italiae cartographica* (1929), pp. 1–13, pl. VII–XIII. There are also studies of individual district maps which sometimes refer to others in the group: A. Bertoldi, 'Topografia del Veronese (secolo XV)', *Archivio Veneto*, 2nd ser. xviii (1888), pp. 455–73; M. Baratta, 'Sopra un'antica carta del territorio bresciano', *Bollettino della Reale Società Geografica*, 5th ser. ii (1913), pp. 514–26, 1025–31, pl. opp. p. 1092; R. Almagià, 'Un'antica carta topografica del territorio veronese', *Rendiconti della Reale Accademia dei Lincei*, clas. sci. mor., xxxii (1925), pp. 63–83; R. Almagià, 'Un'antica carta del territorio di Asti', *Rivista Geografica Italiana*, lviii (1951), pp. 43–4; R. Gallo, 'A Fifteenth-century Military Map of the Venetian Territory of *Terraferma*', *Imago Mundi*, xii (1955), pp. 55–7. The seal map of Frederick II is described and illustrated in G. Schlumberger, F. Chalandon and A. Blanchet, *Sigillographie de l'Orient latin* (1943), p. 22, pl. I.

Of the local maps from medieval Italy that of Talamone is described by W. Braunfels, *Mittelalterliche Stadtbaukunst in der Toskana* (3rd edn, 1966), pp. 77–8, that of the lagoon area around Venice by Almagià, op. cit. (1925), p. 81, those of areas around Ravenna by R. Almagià, *Documenti cartografici dello stato pontificio* (1960), p. 10n. The various versions of Buondelmonte's map of Constantinople are analysed by G. Gerola, 'Le vedute di Costantinopoli di Cristoforo Buondelmonti', *Studi Bizantini e Neo-Ellenici*, iii (1931), pp. 249–79.

4 Town plans and bird's-eye views

Two very different works deal with the whole subject of plans and views of towns in medieval Europe: E. Oberhummer, 'Der Stadtplan, seine Entwickelung und geographische Bedeutung', *Verhandlungen des sechzehnten Deutschen Geographentages zu Nürnberg* (1907), pp. 66–101, and P. Lavedan, *Représentation des villes dans l'art du moyen âge* (1954). The tradition of the city ideogram is discussed in the latter, pp. 33–5, pl. XVII, and also in *Gerasa, City of the Decapolis*, ed. C. H. Kraeling (1938), pp. 341–51.

Arculf's plan of Jerusalem is reproduced in Oberhummer, op. cit., p. 80; the plan of the 1140s with four-sided city wall is discussed by L. H. Heydenreich, 'Ein Jerusalem-Plan aus der Zeit der Kreuzfahrer', *Miscellanea pro arte*, ed. J. Hoster and P. Bloch (1965), pp. 83–90, pl. LXII–LXV; those in the books by Sanudo and Veneto are reproduced and discussed in a general account of the lives and work of these authors by B. Degenhart and A. Schmitt, 'Marino Sanudo und Paolino Veneto', *Römisches Jahrbuch für Kunstgeschichte*, xiv (1973), pp. 78–80, 120–22. Plans of Jerusalem are included in R. Röhricht's work on 'Karten und Pläne zur Palästinakunde aus dem 7. bis 16. Jahrhundert' (see the note dealing with chapter 3), especially in parts 3 and 4.

General lists and reproductions of medieval plans of Rome are in G. B. de Rossi, *Piante iconografiche e prospettiche di Roma anteriori al secolo XVI* (2v., 1879); C. Huelsen, 'Di una nuova pianta prospettica di Roma del secolo XV', *Bullettino della Commissione Archeologica Comunale di Roma*, 4th ser. xx (1892), footnotes on pp. 38–40; and P. Ehrle and H. Egger, *Piante e vedute di Roma e del Vaticano dal 1300 al 1676*, ed. A. P. Frutaz (1956). Studies of particular plans include A. and M. Levi, 'The Medieval Map of Rome in the Ambrosian Library's Manuscript of Solinus', *Proceedings of the American Philosophical Society*, cxviii (1974), pp. 567–94 (fourteenth-century copy of twelfth-century plan); W. Erben, *Rombilder auf kaiserlichen und päpstlichen Siegeln des Mittelalters* (1931), pp. 86–7, 125–7; Degenhart and Schmitt, op. cit., pp. 86–7, 125–7 (plan in Veneto's work); G. Scaglia, 'The Origin of an Archaeological Plan of Rome by Alessandro Strozzi', *Journal of the Warburg and Courtauld Institutes*, xxvii (1964), pp. 137–63; R. Weiss, *The Renaissance Discovery of Classical Antiquity* (1969), pp. 90–94 (Strozzi's and Rosselli's plans).

M. Destombes, 'A Panorama of the Sack of Rome by Pieter Bruegel the Elder', *Imago Mundi*, xiv (1959), pp. 64–73, discusses medieval views and plans of Rome and other Italian cities. Plans of Florence, including the work of Rosselli, are discussed by A. Mori, 'Firenze nelle sue rappresentazioni cartografiche', *Atti della Società Colombaria di Firenze* (1912), pp. 25–42, R. Ciullini, 'Firenze nelle antiche rappresentazioni cartografiche', *Firenze*, ii (1933), pp. 33–7, 129–34, and L. D. Ettlinger, 'A Fifteenth-century View of Florence', *Burlington Magazine*, xciv (1952), pp. 160–67.

J. Schulz, 'Jacopo de' Barbari's View of Venice: Map Making, City Views and Moralized Geography before the year

187

1500', *Art Bulletin*, lx (1978), pp. 425–74, is a fundamentally important study not only of this particular map but of early Italian town plans in general. Of possible measured plans of medieval Italian cities, Schulz, op. cit., pp. 440–41, 445, discusses that of Venice and the quotation by Castiglionchio about one from Florence is taken from Mori, op. cit., p. 30.

The fifteenth-century plan of Vienna is described and analysed by S. Wellisch, 'Der älteste Plan von Wien', *Zeitschrift des Oesterreichischen Ingenieur- und Architekten-Vereines*, l (1898), pp. 757–61, and by M. Kratochwill, 'Zur Frage der Echtheit des "Albertinischen Planes" von Wien', *Jahrbuch des Vereins für Geschichte der Stadt Wien*, xxix (1973), pp. 7–35; that of Bristol by Elizabeth Ralph in *Local Maps and Plans from Medieval England*, ed. R. A. Skelton and P. D. A. Harvey (1980), no. 28. The map of the Holy Land in Breitenbach's work is discussed by R. Röhricht, 'Die Palästinakarte Bernhard von Breitenbachs', *Zeitschrift des Deutschen Palästina-Vereins*, xxiv (1901), pp. 129–35, and by R. Oehme, 'Die Palästinakarte aus Bernhard von Breitenbachs Reise in das Heilige Land', *Aus der Welt des Buches*, lxxv (1951), pp. 70–83. The plan of the universe by Loriti is described and reproduced by W. Blumer, 'Glareanus' Representation of the Universe', *Imago Mundi*, xi (1954), pp. 148–9.

5 The picture-map in medieval Europe

Works relevant to this chapter are best listed region by region. In two cases this can be done very simply. All known British material is presented and discussed in *Local Maps and Plans from Medieval England*, ed. R. A. Skelton and P. D. A. Harvey (1980); the Gloucester survey of 1455, which falls outside its scope, is published as *Rental of all the Houses in Gloucester*, ed. W. H. Stevenson (1890). And from France we have simply the two articles in which F. de Dainville presented his findings: *Ecole Pratique des Hautes Etudes, IVe section: sciences historiques et philologiques. Annuaire 1968–1969* (1969), pp. 397–408, and 'Cartes et contestations au XVe siècle', *Imago Mundi*, xxiv (1970), pp. 99–121 (the quotation by Zwolle is taken from p. 109). The fullest edition of Honnecourt's notebook with its plans of buildings is *Villard de Honnecourt: kritische Gesamtausgabe des Bauhüttenbuches*, ed. H. R. Hahnloser (2nd edn, 1972).

On medieval local maps from the Low Countries we have some general discussion by, particularly, J. Keuning, 'Sixteenth Century Cartography in the Netherlands (Mainly in the Northern Provinces)', *Imago Mundi*, ix (1952), pp. 40–42, B. van't Hoff, 'The Oldest Maps of the Netherlands: Dutch Map Fragments of about 1524', *Imago Mundi*, xvi (1962), pp. 29–32, and C. Koeman, 'Algemene inleiding over de historische kartografie,

meer in het bijzonder: Holland vóór 1600', *Holland*, vii (1975), pp. 218–37. But mostly we have detailed studies only of individual maps or small groups, among them D. T. Enklaar, 'De oudste kaarten van Gooiland en zijn grensgebieden', *Nederlandsch Archievenblad*, xxxix (1931–2), pp. 185–205; M. K. E. Gottschalk, 'De oudste kartografische weergave van een deel van Zeeuwsch-Vlaanderen', *Archief* (1948), pp. 30–39; M. K. E. Gottschalk and W. S. Unger, 'De oudste kaarten der waterwegen tussen Brabant, Vlaanderen en Zeeland', *Tijdschrift van het Koninklijk Nederlandsch Aardrijkskundig Genootschap*, 2nd ser. lxvii (1950), pp. 146–64; M. K. E. Gottschalk, *Historische geografie van westelijk Zeeuws-Vlaanderen* (2v., 1955–8), i, pp. 148–9 (the 1307 plan of an area near Sluis); A. H. Huussen, *Jurisprudentie en kartografie in de XVe en XVIe eeuw* (1974) (maps in the archives of the Mechelen Great Council). Interesting information on surveying in the medieval Low Countries is provided by P. S. Teeling's series of articles on 'Oud-Nederlandse landmeters' in *Orgaan der Vereeniging van Technische Ambtenaren van het Kadaster*, vii and viii (1949–50), and by A. Viaene, 'De landmeter in Vlaanderen 1281–1800', *Biekorf*, lxvii (1966), pp. 1–19.

Medieval local maps from south-west Germany are mentioned by R. Oehme, *Die Geschichte der Kartographie des deutschen Südwestens* (1961), pp. 14, 85, 97; the map of Wantzenau is described by F. Grenacher, 'Current Knowledge of Alsatian Cartography', *Imago Mundi*, xviii (1964), pp. 60–61, those of Beringsweiler by K. Schumm, *Inventar der handschriftlichen Karten im Hohenlohe-Zentralarchiv Neuenstein* (1961), p. 5. A good deal has been written about the plan of Lake Constance by 'PW'; W. Bonacker, 'Die sogenannte Bodenseekarte des Meisters PW bzw. PPW vom Jahre 1505', *Die Erde*, lxxxv (1954), pp. 1–29, gathers together the varying opinions of earlier writers. The building-plans from Vienna are reproduced and discussed by H. Koepf, *Die gotischen Planrisse der Wiener Sammlung* (1969).

The maps from medieval Poland are described by K. Buczek, *The History of Polish Cartography from the 15th to the 18th Century*, ed. A. Potocki (1966), pp. 22–4; the two sketch maps from Pomerania are reproduced and discussed in detail by B. Olszewicz, *Dwie szkicowe mapy Pomorza z połowy XV wieku* (1937).

The Portuguese book of fortresses by De Armas is described, with reproductions, in *Portugaliae monumenta cartographica*, ed. A. Cortesão and A. Teixeira da Mota (6v., 1960–62), i, pp. 71–5, pl 28–33.

6 The picture-map in the Far East

All previous western work on Chinese cartography has been effectively superseded by the authoritative account of J.

Needham, *Science and Civilisation in China* (in progress, 1954–), iii, pp. 497–590; the quotation from Phei Hsiu is taken from p. 539. However, two earlier articles dealing particularly with the Ming and Manchu dynasties may be mentioned – both are in *Monumenta Serica*, i (1935–6): W. Fuchs, 'Materialien zur Kartographie der Mandzu-Zeit' (pp. 386–427) and H. Bernard, 'Les étapes de la cartographie scientifique pour la Chine et les pays voisins' (pp. 428–77). The Early Han dynasty maps from the Changsha area are discussed by M. Hsu, 'The Han Maps of the Second Century B.C.: their Quality and Historical Importance', an unpublished paper delivered at the 7th International Conference on the History of Cartography (Washington, 1977), but they are also described and reproduced in *China Pictorial*, 1974 no. 11, pp. 36–41, and 1975 no. 9, pp. 34–7, and in *Wen Wu (Cultural Relics)*, i (1976), pp. 18–23. Accounts of particular picture-maps from China include J. V. Mills, 'Chinese Coastal Maps', *Imago Mundi*, xi (1954), pp. 151–68, and M. J. Meijer, 'A Map of the Great Wall of China', *Imago Mundi*, xiii (1956), pp. 110–15. Chinese mapping in eighteenth-century Tibet can be demonstrated from W. W. Rockhill, 'Tibet: a Geographical, Ethnographical and Historical Sketch, Derived from Chinese Sources', *Journal of the Royal Asiatic Society* (1891), maps before p. 1 and opp. pp. 70, 185, and from R. H. Phillimore, *Historical Records of the Survey of India*, i (1945), p. 70, and in nineteenth-century Mongolia from M. Venyukof, 'New Maps of Mongolia', *Geographical Magazine*, iii (1876), pp. 127–8.

There is no western-language history of topographical mapping in Japan. Two general works on Japanese cartography are almost entirely concerned with geographic maps: P. Teleki, *Atlas zur Geschichte der Kartographie der japanischen Inseln* (1909), and E. W. Dahlgren, *Les débuts de la cartographie du Japon* (1911). However, a recent work, with Japanese text but English captions, presents a magnificent conspectus of topographical as well as geographic maps from Japan in colour reproductions: M. Nanba and K. Unno, *Nihon no ko chizu (Old Maps in Japan)* (1970). M. Ramming, 'The Evolution of Cartography in Japan', *Imago Mundi*, ii (1937), pp. 17–21, gives a brief account of the very earliest Japanese maps; the later tradition of plans of towns is discussed in detail by M. Kurita, 'Japanese Old Printed Maps of Cities', *Comptes rendus du Congrès International de Géographie: Amsterdam 1938* (2v., 1938), ii, Travaux sec. A–F, pp. 362–80, and a well illustrated catalogue of those from one city is provided by M. P. McGovern, 'A List of Nagasaki Maps Printed during the Tokugawa Era', *Imago Mundi*, xv (1960), pp. 105–10. L. Serrurier, *Bibliothèque japonaise: catalogue* raisonné des livres et des manuscrits japonais enregistrés à la bibliothèque de l'université de Leyde (1896), pp. 43–94, gives a catalogue of the maps from Japan in one of the principal European collections, with helpful descriptive notes and comments.

Maps from Burma are described and reproduced in four articles by F. Hamilton in the Edinburgh Philosophical Journal: 'An Account of a Map of the Countries Subject to the King of Ava, Drawn by a Slave of the King's Eldest Son' (ii (1820), pp. 89–95, 262–71, pl. X; the quotation is from pp. 89–90), 'Account of a Map of the Country North from Ava' (iv (1820–21), pp. 76–87, pl. II), 'Account of a Map of the Country between the Erawadi and Khiaenduaen Rivers' (vi (1822), pp. 107–11, pl. IV), and 'Account of a Map Drawn by a Native of Dawae or Tavay' (ix (1823), pp. 228–36, pl. V). The fifteenth-century map from western Java is discussed by R. Kusmiadi, 'A Brief History on Cartography in Indonesia', an unpublished paper delivered at the 7th International Conference on the History of Cartography (Washington, 1977).

7 Mexico and India

Brief general accounts of early Mexican maps are given by M. Orozco y Berra, Materiales para una cartografia mexicana (1871), pp. 1–13, Eulalia Guzmán, 'The Art of Map-making among the Ancient Mexicans', Imago Mundi, iii (1939), pp. 1–6, and C. A. Burland, 'The Map as a Vehicle of Mexican History', Imago Mundi, xv (1960), pp. 11–18. Interesting analyses of two particular maps which both contain a mixture of Aztec and Spanish elements are in B. Díaz del Castillo, The True History of the Conquest of New Spain, ed. A. P. Maudslay (Hakluyt Soc., 5v., 1908–16), iii, pp. 1–25 with facsimile in four sheets (map of western suburbs of the city of Mexico), and H. F. Cline, 'The Oztoticpac Lands Map of Texcoco, 1540', Quarterly Journal of the Library of Congress, xxiii (1966), pp. 76–115 (where, inter alia, plans of buildings are discussed). The quotations from the letters of Cortés are from Fernando Cortes: his Five Letters of Relation to the Emperor Charles V, ed. F. A. MacNutt (2v., 1908), i, pp. 244–5, and ii, pp. 231–2, and that from Díaz del Castillo is from The True History, v, p. 12.

There is no general account or discussion of pre-European traditions of map-making in India, but three notes that at least draw attention to the problem are by R. H. Phillimore, 'Indigenous Indian Maps', The Indian Archives, iv (1950), pp. 43–4, D. C. Sircar, 'Ancient Indian Cartography', The Indian Archives, v (1951), pp. 60–63, and R. L. Singh, L. R. Singh and B. Dube, 'The Ancient Indian Contribution to Cartography', National Geographical Journal of India, xii (1966), pp. 24–37. Reproductions and descriptions of individual maps are given by R. H. Phil-

limore, 'Three Indian Maps', Imago Mundi, ix (1952), pp. 111–14, and C. D. Deshpande, 'A Note on Maratha Cartography', The Indian Archives, vii (1953), pp. 87–94. References to maps from Nepal are in Narratives of the Mission of George Bogle to Tibet and of the Journey of Thomas Manning to Lhasa, ed. C. R. Markham (1876), p. cxxvii (cf. p. lxxixn), and in H. de Hutorowicz, 'Maps of Primitive Peoples', Bulletin of the American Geographical Society, xliii (1911), p. 676. The quotation by Wilford is from 'An Essay on the Sacred Isles in the West', Asiatick Researches, viii (1805), p. 271.

8 The earliest scale-maps

There is a short general survey of the cartography of ancient Mesopotamia by E. Unger, 'Ancient Babylonian Maps and Plans', Antiquity, ix (1935), pp. 311–22. The full analysis of building-plans by E. Heinrich and Ursula Seidl is 'Grundrisszeichnungen aus dem alten Orient', Mitteilungen der Deutschen Orient-Gesellschaft zu Berlin, xcviii (1967), pp. 24–45. Other descriptions and discussions of individual plans include F. Thureau-Dangin, 'Un cadastre chaldéen', Revue d'Assyriologie, iv (1898), pp. 13–27 (plans of fields, buildings, waterways), L. W. King, Babylonian Boundary-stones and Memorial-tablets in the British Museum (2v., 1912), pp. 81–2, 85–6, 99–101, pl. 14, 16, 17 (plans on kudurru), S. H. Langdon, 'An Ancient Babylonian Map', The Museum Journal, vii (1916), pp. 263–8 (plan of fields near Nippur), E. D. van Buren, Clay Figurines of Babylonia and Assyria (1930), pp. 274–6 (plans of cities, fields, buildings).

Evidence for surveying and mensuration in ancient Egypt is brought together and discussed by A. Déléage, 'Les cadastres antiques jusqu'à Dioclétien', Etudes de Papyrologie, ii (1934), pp. 73–228; the early measurement of land there is also described by T. E. Peet, The Rhind Mathematical Papyrus (1923), pp. 3, 6, 9, 24–5, 31–2, 88, and the use of scale drawings by F. Petrie, 'Egyptian Working Drawings', Ancient Egypt (1926), part 1, pp. 24–7.

There is an account of the work of surveyors in the Roman Empire in the same article by Déléage; another, more recent, is by O. A. W. Dilke, The Roman Land Surveyors (1971). The latter includes a chapter on the Orange surveys, but the fullest account of these is by A. Piganiol, Les documents cadastraux de la colonie romaine d'Orange (Gallia supplement 16, 1962). All known fragments of the third-century plan of Rome are published, together with those from other similar Roman scale-maps and a very full discussion, in G. Carettoni, A. M. Colini, L. Cozza and G. Gatti, La pianta marmorea di Roma antica: Forma urbis Romae (1960). Arculf's plans of buildings in the Holy Land are reproduced, and surviving manuscripts listed, in Itinera Hierosolym-

itana et descriptiones terrae sanctae, ed. T. Tobler and A. Molinier (Soc. de l'Orient Latin, 1879), pp. xxx–xxxiii, 149, 160, 165, 181. On the monastic plan from St Gall a great deal has been written, much of it controversial; a simple introduction is provided by H. Reinhardt, Der St. Galler Klosterplan (1952), while W. Horn and E. Born, 'New Theses about the Plan of St. Gall', in Die Abtei Reichenau: neue Beiträge zur Geschichte und Kultur des Inselklosters, ed. H. Maurer (1974), pp. 407–76, survey the more recent writings pending the appearance of the same authors' definitive work on the plan.

9 From itinerary to survey

Chinese scale-maps are described and discussed in J. Needham, Science and Civilisation in China (in progress, 1954–   ), iii, pp. 497–590, especially pp. 533–56; the quotations from Phei Hsiu and the preface to Chu Ssu-pên's atlas are taken from pp. 539–40, 552.

The quotation from Vegetius is from his Epitoma rei militaris, book III, chapter 6 (ed. C. Lang (1885), p. 75). The Antonine and Bordeaux Itineraries are published in Itineraria romana: Itineraria Antonini Augusti et Burdigalense, ed. O. Cuntz (1929). The Rudge Cup is illustrated and discussed by J. D. Cowen and I. A. Richmond, 'The Rudge Cup', Archaeologia Aeliana, 4th ser. xii (1935), pp. 310–42, pl. XXVIII, the Dura-Europos shield by F. Cumont, Fouilles de Doura-Europos (1926), pp. 323–37, pl. CIX. A full coloured reproduction of the Peutinger Table has been published, with an introduction by E. Weber, as Tabula Peutingeriana: Codex Vindobonensis 324 (2v., 1976); A. and M. Levi, Itineraria picta: contributo allo studio della Tabula Peutingeriana (Museo dell'Impero Romano, 1967), discuss in detail its form and style and its relationship to other Roman maps.

The Bruges Itinerary is analysed by J. Lelewel, Géographie du moyen âge (4v. and epilogue, 1852–7), epilogue, pp. 281–308, the itinerary of Sigeric by F. P. Magoun, 'An English Pilgrim-diary of the Year 990', Mediaeval Studies, ii (1940), pp. 231–52, the Titchfield itinerary by B. Dickins, 'Premonstratensian Itineraries from a Titchfield Abbey Manuscript at Welbeck', Proceedings of the Leeds Philosophical and Literary Society: Literary and Historical Section, iv (1936–8), pp. 349–61. The maps of Wormley and Sherwood Forest are discussed, by P. D. A. Harvey and M. W. Barley respectively, in Local Maps and Plans from Medieval England, ed. R. A. Skelton and P. D. A. Harvey (1980), nos 2, 10. Matthew Paris's itineraries to Apulia are reproduced and discussed, together with his other maps, in R. Vaughan, Matthew Paris (1958), pp. 235–50, pl. XII, XIII, XV; all four versions of his map of Great Britain are reproduced in colour in Four Maps of Great Britain Designed by Mat-

thew Paris (British Museum, 1928) and are also discussed by J. B. Mitchell, 'Early Maps of Great Britain: The Matthew Paris Maps', *Geographical Journal*, lxxxi (Jan.–June 1933), pp. 27–34. The Gough Map has been reproduced in colour as *The Map of Great Britain, circa A.D. 1360, Known as the Gough Map* (Royal Geographical Soc. and Bodleian Library, 1958) accompanied by a memoir by E. J. S. Parsons which discusses its date, its content and its method of compilation.

The map of the Holy Land in Marino Sanudo's *Liber secretorum* is discussed by R. Röhricht, 'Marino Sanudo sen. als Kartograph Palästinas', *Zeitschrift des Deutschen Palästina-Vereins*, xxi (1898), pp. 84–126, pl. 2–11, in great detail, as well as by Needham, op. cit., iii, p. 564, pl. LXXXVII, LXXXVIII, and by B. Degenhart and A. Schmitt, 'Marino Sanudo und Paolino Veneto', *Römisches Jahrbuch für Kunstgeschichte*, xiv (1973), pp. 76–8, 116–19. For the district maps from north Italy and the Naples boundary map see the note dealing with chapter 3.

The maps of Cusa and Etzlaub are discussed by A. Wolkenhauer, 'Über die ältesten Reisekarten von Deutschland aus dem Ende des 15. und dem Anfange des 16. Jahrhunderts', *Deutsche Geographische Blätter*, xxvi (1903), pp. 120–38. The quotation on Cusa's maps is from E. Lehmann, *Alte deutsche Landkarten* (1935), p. 31. More recent work on Etzlaub includes H. Krüger, 'Erhard Etzlaub's *Romweg* Map and its Dating in the Holy Year of 1500', *Imago Mundi*, viii (1951), pp. 17–26, F. Schnelbögl, 'Life and Work of the Nuremberg Cartographer Erhard Etzlaub', *Imago Mundi*, xx (1966), pp. 11–26, and T. Campbell, 'The Woodcut Map as a Physical Object: a New Look at Erhard Etzlaub's "Rom Weg" Map of c. 1500', unpublished paper at the 7th International Conference on the History of Cartography (Washington, 1977). On the use of the compass in map-making see Needham, op. cit., iii, pp. 576–7, and Skelton and Harvey, op. cit. Türst's map of Switzerland is reproduced and discussed by E. Imhof, *Die ältesten Schweizerkarten* (1939), pp. 6–14, pl. 1, 2, and (a very full history and analysis) by T. Ischer, *Die ältesten Karten der Eidgenossenschaft* (1945).

10 Sixteenth-century Europe

The maps by Da Vinci in the Royal Collection are superbly reproduced and fully discussed by M. Baratta, *I disegni geografici di Leonardo da Vinci conservati nell castello di Windsor* (1941); an earlier discussion of Da Vinci's cartography, referring also to the work of Dürer and other contemporary artists, is by E. Oberhummer, 'Leonardo da Vinci and the Art of the Renaissance in its Relations to Geography', *Geographical Journal*, xxxiii (Jan.–June 1909), pp. 540–69.

The quotation from Beale's treatise is taken from C. Read, *Mr Secretary Walsingham* (3v., 1925), i, p. 428. Burghley's concern for maps is demonstrated by R. A. Skelton and J. N. Summerson, *A Description of Maps and Architectural Drawings in the Collection made by William Cecil First Baron Burghley now at Hatfield House* (Roxburghe Club, 1971), Hakluyt's unconcern by R. A. Skelton, 'Hakluyt's Maps', *The Hakluyt Handbook* (Hakluyt Soc., 1974), pp. 48–73. The sixteenth-century manuscript picture-maps that are mentioned, apart from those actually illustrated, are reproduced in *Buckinghamshire Estate Maps* (Buckinghamshire Record Soc., 1964), pl. 2 (Brill, 1590), S. A. Moore and P. Birkett, *A Short History of the Rights of Common upon the Forest of Dartmoor* (1890), between pp. 166, 167 (Dartmoor, c. 1540) and R. Oehme, *Die Geschichte der Kartographie des deutschen Südwestens* (1961), opp. p. 24 (Fils valley, c. 1534). The tradition of picture-maps in Franconia is described by W. M. Brod, 'Fränkische Hof- und Stadtmaler als Kartographen', *Kartengeschichte und Kartenbearbeitung*, ed. K. H. Meine (1968), pp. 49–57. The two plates of the map of London in the 1550s are reproduced and discussed by M. Holmes, 'An Unrecorded Map of London', *Archaeologia*, c (1966), pp. 105–28.

The development of the scale-map and surveying techniques in contrasting areas of Germany can be seen in J. Werner, *Die Entwicklung der Kartographie Südbadens im 16. und 17. Jahrhundert* (1913), and Oehme, op. cit., on the one hand, and in F. Geerz, *Geschichte der geographischen Vermessungen und der Landkarten Nordalbingiens vom Ende des 15. Jahrhunderts bis zum Jahre 1859* (1859), which deals with Schleswig-Holstein (with some reference also to Danish cartography). For Etzlaub see the note dealing with chapter 9. The construction of Brahé's map of Hven is discussed by H. Richter, 'Willem Jansz. Blaeu with Tycho Brahe on Hven and his Map of the Island: Some New Facts', *Imago Mundi*, iii (1939), pp. 53–60.

Useful guides to early manuals on surveying in Britain are provided by H. C. Darby, 'The Agrarian Contribution to Surveying in England', *Geographical Journal*, lxxxii (July-Dec. 1933), pp. 529–35, and by A. W. Richeson, *English Land Measuring to 1800: Instruments and Practices* (Soc. for the History of Technology, 1966). D. Hodson's work on the early mapping of Portsmouth is published as *Maps of Portsmouth before 1801* (Portsmouth Record Series, 1978). The contribution of military engineers to the spread in Britain of accurate surveying and scale-maps is discussed in Skelton and Summerson, op. cit., and also by R. A. Skelton, 'The Military Surveyor's Contribution to British Cartography in the Sixteenth Century', *Imago Mundi*, xxiv (1970), pp. 77–83. The sixteenth-

century King's College surveys without maps are described by W. J. Corbett, 'Elizabethan Village Surveys', *Transactions of the Royal Historical Society*, 2nd ser. xi (1897), pp. 67–87, while several of the All Souls College maps are reproduced in M. W. Beresford, *History on the Ground* (1957). The quotation from Ralph Agas is from British Library, Lansdowne MS. 165, f. 91.

General surveys of cartography in the sixteenth-century Low Countries are provided by S. J. Fockema Andreae, *Geschiedenis der kartografie van Nederland van den romeinischen tijd tot het midden der 19de eeuw* (1947) and J. Keuning, 'Sixteenth Century Cartography in the Netherlands (Mainly in the Northern Provinces)', *Imago Mundi*, ix (1952), pp. 35–63. The series of articles by P. S. Teeling on 'Oud-Nederlandse landmeters' in *Orgaan der Vereeniging van Technische Ambtenaren van het Kadaster*, vii and viii (1949–50), presents valuable material on developments in surveying, especially on the 'little foot' (vii, pp. 160–61) and the work of Van Deventer (vii, pp. 163–6). Also relevant to Van Deventer are F. C. Wieder, *Nederlandsche historisch-geographische documenten in Spanje* (Koninklijk Nederlandsch Aardrijkskundig Genootschap, 1915), and B. van't Hoff, 'The Oldest Maps of the Netherlands: Dutch Map Fragments of about 1524', *Imago Mundi*, xvi (1962), pp. 29–32. Other works on particular map-makers – either individuals or groups – in the sixteenth-century Low Countries include H. E. Wauwermans, *Histoire de l'école cartographique belge et anversoise du XVIe siècle* (2v., 1895); J. Denucé, *Oud-Nederlandsche kaartmakers in betrekking met Plantijn* (Maatschapij der Antwerpsche Bibliophilen, 2v., 1912–13); A. de Smet, 'La cartographie scientifique à Louvain de 1500 à 1550', in *Kartengeschichte und Kartenbearbeitung*, ed. K. H. Meine (1968), pp. 59–61; A. H. Huussen, 'Willem Hendricxz. Croock, Amsterdams stadsfabriekmeester, schilder en kartograaf in de eerste helft van de zestiende eeuw', *Jaarboek van het Genootschap Amstelodamum*, lxiv (1972), pp. 29–53; P. Ratsma, 'De landmeter Jan Jansz. Potter: de topografie van Rotterdam en omgeving in de tweede helft van de zestiende eeuw', *Holland*, vii (1975), pp. 300–21.

11 The pictorial inheritance

The themes of this chapter can be illustrated from practically any history of cartography or collection of reproductions of maps. Notes on ornament and conventional signs, with particular reference to British maps, are by E. Lynam, 'The Development of Symbols, Lettering, Ornament and Colour on English Maps', *British Records Association: Proceedings*, iv (1939), pp. 20–34; E. Lynam, 'Period Ornament, Writing and Symbols on Maps, 1250–1800', *Geographical Mag-*

azine, xviii (1946), pp. 365–8, fig. 19–37; and E. M. J. Campbell, 'Landmarks in British Cartography: the Beginnings of the Characteristic Sheet to English Maps', *Geographical Journal*, cxxviii (1962), pp. 411–15. Rocque's Kildare estate survey and its draft vignette are described and reproduced by H. Cobbe, 'Four Manuscript Maps Recently Acquired by the British Museum', *Journal of the Society of Archivists*, iv, no. 8 (Oct. 1973), pp. 650–52, pl. II; the work of Hornor is fully described and illustrated by R. Hyde, 'Thomas Hornor: Pictural Land Surveyor', *Imago Mundi*, xxix (1977), pp. 23–34.

The development of hachuring and contour lines is described by K. Peucker, 'Zur kartographischen Darstellung der dritten Dimension', *Geographische Zeitschrift*, vii (1901), pp. 22–41, by J. Röger, *Die Geländedarstellung auf Karten* (1908) and by F. de Dainville, 'De la profondeur à l'altitude: des origines marines de l'expansion cartographique du relief terrestre par cotes et courbes de niveau (XVIIe–XVIIIe siècles)', *Le navire et l'économie maritime*, ed. M. Mollat (Ecole Pratique des Hautes Etudes, VIe Section, 1958), pp. 195–213. Evolving methods of showing relief can be clearly seen in the illustrations to F. Massie, *La cartographie des Pyrénées* (Section des Hautes Pyrénées du Club Alpin Français, 1934), and to W. Blumer, *Bibliographie der Gesamtkarten der Schweiz von Anfang bis 1802* (1957); the quotation by H. Béraldi is from his *Balaïtous et Pelvoux*, i (1907), p. 37, as cited by Massie, op. cit., p. 26.

List of illustrations

192

Index

Numbers in italics refer to illustrations